FREELIVING MARINE NEMATODES

Part I. BRITISH ENOPLIDS

A NEW SERIES

Synopses of the British Fauna

Edited by Doris M. Kermack and R. S. K. Barnes

The *Synopses of the British Fauna* are illustrated field and laboratory pocket-books designed to meet the needs of amateur and professional naturalists from sixth-form level upwards. Each volume presents a more detailed account of a group of animals than is found in most field-guides and bridges the gap between the popular guide and more specialist monographs and treatises. Technical terms are kept to a minimum and the books are therefore intelligible to readers with no previous knowledge of the group concerned.

Volumes 1–18 inclusive are available only from the Linnean Society of London

1. *British Ascidians* R. H. Millar
2. *British Prosobranchs* Alastair Graham
3. *British Marine Isopods* E. Naylor
4. *British Harvestmen* J. H. P. Sankey and T. H. Savory
5. *British Sea Spiders* P. E. King
6. *British Land Snails* R. A. D. Cameron and Margaret Redfern
7. *British Cumaceans* N. S. Jones
8. *British Opisthobranch Molluscs* T. E. Thompson and Gregory H. Brown
9. *British Tardigrades* C. I. Morgan and P. E. King
10. *British Anascan Bryozoans* J. S. Ryland and P. J. Hayward
11. *British Freshwater Bivalve Mollusca* A. E. Ellis
12. *British Sipunculans* P. E. Gibbs
13. *British and Other Phoronids* C. C. Emig
14. *British Ascophoran Bryozoans* P. J. Hayward and J. S. Ryland
15. *British Coastal Shrimps and Prawns* G. Smaldon
16. *British Nearshore Foraminiferids* John W. Murray
17. *British Brachiopods* C. H. C. Brunton and G. B. Curry
18. *British Anthozoans* R. L. Manuel
19. *British Planarians* I. R. Ball and T. B. Reynoldson
20. *British Pelagic Tunicates* J. H. Fraser
21. *British and Other Marine and Estuarine Oligochaetes* R. O. Brinkhurst
22. *British and Other Freshwater Ciliated Protozoa: Part I* C. R. Curds
23. *British and Other Freshwater Ciliated Protozoa: Part II* C. R. Curds, M. A. Gates and D. McL. Roberts
24. *British Nemerteans* R. Gibson
25. *Shallow-water Crabs* R. W. Ingle
26. *Polyclad Turbellarians* S. Prudhoe
27. *Tanaids* D. M. Holdich and J. A. Jones

Futher titles are in preparation

No. 28

FREELIVING MARINE NEMATODES

Part I

BRITISH ENOPLIDS

Pictorial key to world genera and notes for the identification of British species

HOWARD M. PLATT

British Museum (Natural History), London SW7 5BD

and

RICHARD M. WARWICK

N.E.R.C. Institute for Marine Environmental Research, Prospect Place, The Hoe, Plymouth PL1 3DH

1983

Published for

The Linnean Society of London

and

The Estuarine and Brackish-water Sciences Association

by

Cambridge University Press

Cambridge

London New York New Rochelle

Melbourne Sydney

Published by the Press Syndicate of the University of Cambridge
The Pitt Building, Trumpington Street, Cambridge CB2 1RP
32 East 57th Street, New York, NY 10022, USA
296 Beaconsfield Parade, Middle Park, Melbourne 3206, Australia

First published 1983

Printed in Great Britain at The Pitman Press, Bath

Library of Congress catalogue card number: 82-19838

British Library Cataloguing in Publication Data
Free-living marine nematodes. – (Synopses of the British fauna; 28)
Part 1: British enoplids
1. Nematoda
I. Platt, H.M. II. Warwick, R.M. III. Series 595.1'82 QL39.N4

ISBN 0 521 25422 1

A Synopsis of the Freeliving Marine Nematodes Part I. British Enoplids

HOWARD M. PLATT

British Museum (Natural History), Cromwell Road, London SW7 5BD

and

RICHARD M. WARWICK

N.E.R.C. Institute for Marine Environmental Research, Prospect Place, The Hoe, Plymouth PL1 3DH

Contents

Foreword

To most marine biologists the nematode component of their samples is an embarrassment and it often appears as 'unidentified worms' in survey reports. This *Synopsis* is the first of three which will be devoted to the marine members of this phylum and describes the nematodes belonging to the subclass Enoplia, so that these marine worms can be given the dignity of a name in field reports in the future. At the beginning of this *Synopsis* is a section dealing with the structure and biology of marine nematodes generally. This is followed by a pictorial key to help identify the different worldwide genera. Following usual editorial policy this introductory part and the pictorial key will be repeated with suitable modifications in the *Synopses* devoted to chromadorids and the monhysterids, so that each is complete in itself.

This is the first time that wholly pictorial keys combining several characters and character-combinations have been used in a synopsis. The Editors will be interested to hear how easy they are to use and if they are arranged in the best possible way upon the pages devoted to them. If readers could let the Editors have their comments within the next year or so it may be possible to incorporate them in further nematode *Synopses*.

The Authors state in their Introduction 'Any investigation of a new area is almost certain to encounter forms presently unknown from British waters and possibly new to science'. The detailed study of freeliving nematodes is clearly in its infancy and awaits workers, both with and without diving gear, to investigate these small but very abundant worms and to expand our knowledge of them. The Editors thank the Authors for the imaginative way in which they have presented their knowledge of this little understood group and sincerely hope that it will stimulate further investigations. We all anxiously await the appearance of the other *Synopses*, devoted to the chromadorid and monhysterid nematodes.

R. S. K. Barnes
Estuarine and Brackish-water
Sciences Association

Doris M. Kermack
The Linnean Society
of London

General introduction to marine nematodes

The Phylum Nematoda consists of small multicellular vermiform organisms which can be found in almost every conceivable environment. Apart from existing as freeliving forms in soil, freshwater and marine benthic habitats, they also parasitise plants and other animals, including man. The whole phylum currently contains some 20,000 nominal species. About 4000 species are freeliving marine forms and, of these, some 450 representing 154 genera have so far been reported from the British Isles (here taken to include Ireland).

Although these synopses are designed to be practical aids to the collection and identification of species from Britain and Ireland, this first volume on nematodes contains a guide to marine genera worldwide. In the past, dichotomous keys for marine nematodes have, for a number of reasons, proved to be unworkable. It has therefore been decided to adopt a more novel approach using pictorial keys which enable the user to evaluate visually several characters and character combinations simultaneously. The inclusion of all known genera is necessary because nematodes from only a limited range of British habitats and localities have been investigated: the number of genera and species so far recorded is only a small fraction of the probable total. Any investigation of a new area is almost certain to encounter forms presently unknown from British waters and possibly new to science. Therefore, a treatment of the known British genera alone would necessarily be incomplete. Separate volumes will deal with the three major taxonomic groups, the enoplids, chromadorids and monhysterids, and include descriptions of British species. As in the past where a group of animals is covered by more than one synopsis, e.g. Cheilostomata (*Synopses 10* and *14*) and Freshwater Ciliated Protozoa (*Synopses 22* and *23*), the general introduction and pictorial keys will be repeated with suitable modifications, so each volume is complete in itself.

Structure

At the microscopic level, marine nematodes are quite varied morphologically and no one species can be considered representative. Therefore, the following description of the general structure is illustrated with a generalised nematode (Fig. 1).

Most adult marine nematodes are elongated cylindrical worms around 1–2 mm long, although some of the larger forms such as those found in kelp holdfasts may be several millimetres in length. Fortunately, the vast majority are sufficiently transparent to allow their internal anatomy to be seen without recourse to special methods of preparation and this considerably increases the number of characters available for practical identification purposes.

The **body** is essentially a tube within a tube. The external tube is the body wall consisting externally of a **cuticle layer** and internally of a **longitudinal muscle layer**. Nematodes do not possess circular musculature so are unable to elongate the body. Movement is brought about by alternate contractions of the dorsal and ventral muscle blocks (Fig. 1C, D) working against the hydrostatic skeleton provided by having a high internal turgor pressure. This results in the highly characteristic serpentine pattern of locomotion. Because the flexing is in the dorso-ventral plane, the fixed nematode normally comes to lie on a microscope slide with a lateral side uppermost, which is why most illustrations are of lateral views.

The internal tube is the **gut** which is terminal at the anterior but subterminal posteriorly, so that nematodes have a **tail**. The gut is differentiated into a **buccal cavity**, a muscular **oesophagus** (more accurately termed a **pharynx**), an **intestine** and a short **rectum**. Between gut and body wall lies the fluid-filled cavity called a **pseudocoelom** (so-called because it appears to lack an epithelial lining) in which the reproductive organs are to be found.

The body wall and gut provide many characters used in identification and, together with taxonomically useful parts of the nervous, excretory and reproductive systems, they will now be described in more detail.

Cuticle

The **cuticle** may appear to be entirely smooth or to be transversely annulated, the latter usually being called a **striated cuticle**. In many groups, there may

Fig. 1. General features. A, Generalised male marine nematode, lateral view. B, Generalised female marine nematode, lateral view. C, Transverse section through oesophageal region. D, Transverse section through intestinal region.

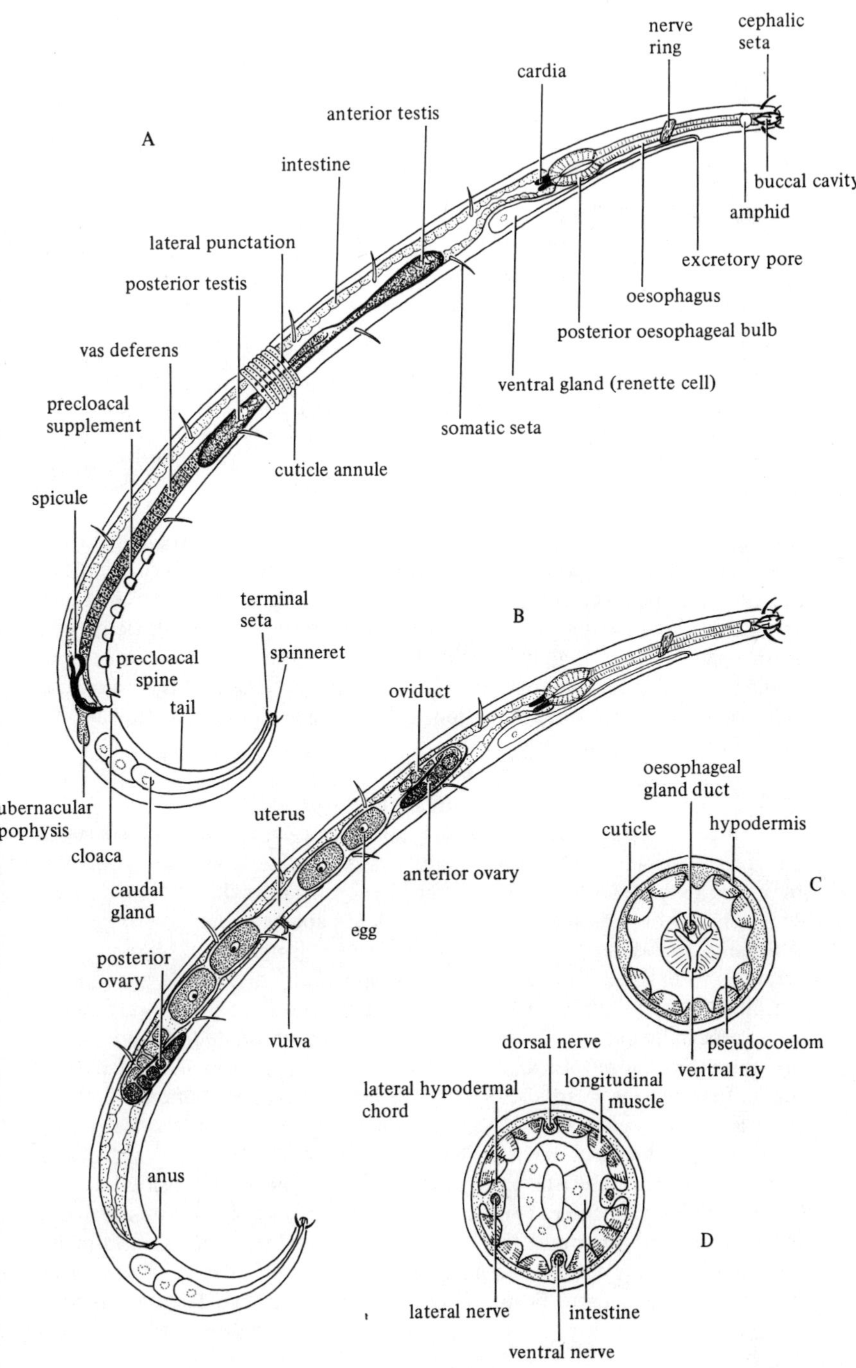
A
cephalic seta
nerve ring
cardia
anterior testis
intestine
lateral punctation
posterior testis
vas deferens
precloacal supplement
spicule
buccal cavity
amphid
excretory pore
oesophagus
posterior oesophageal bulb
ventral gland (renette cell)
somatic seta
cuticle annule
terminal seta
spinneret
precloacal spine
tail
gubernacular apophysis
cloaca
caudal gland
B
oviduct
uterus
anterior ovary
egg
posterior ovary
vulva
anus
oesophageal gland duct
cuticle
hypodermis
C
pseudocoelom
ventral ray
dorsal nerve
lateral hypodermal chord
longitudinal muscle
D
lateral nerve
intestine
ventral nerve

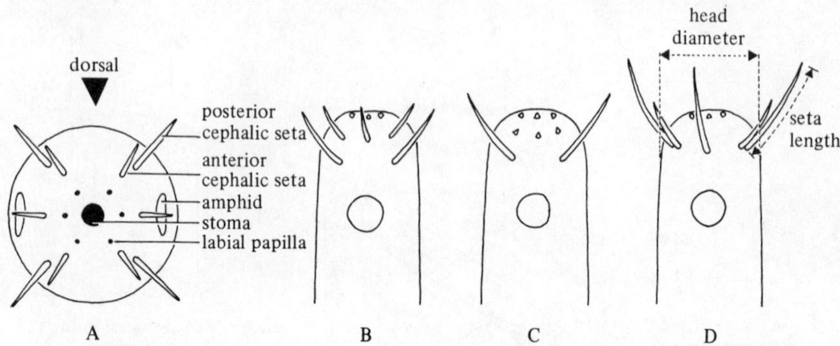

Fig. 2. Head sensilla. A, Apical view showing typical distribution of head sensilla. B, Head sensilla in the 6 + 6 + 4 pattern. C, Head sensilla pattern where only the third circle are setiform. D, Head sensilla in the 6 + 10 pattern, also showing setal measurements.

also appear to be dots (**punctations**) which can be arranged irregularly or in rows. These and various other elaborations of the cuticle may be confined to the lateral parts of the cuticle only, when it is termed **lateral differentiation**. Examples of cuticle patterns are shown in Fig. 15. Where the cuticle pattern is the same in all regions of the body it is sometimes referred to as homogeneous; where the pattern changes along the length of the body, it is called heterogeneous ornamentation. In addition to dots, there may also be pores in the cuticle laterally.

Sense organs

The cuticle bears a number of sensory structures, collectively called **sensilla** (singular: sensillum) which, although having an underlying basic structure, may take a variety of forms. Long hair-like sensilla are called **setae** (singular: seta) while if only nipple-like they are called **papillae** (singular: papilla).

Sensilla found on the general body surface (**somatic** sensilla) may be arranged in definite longitudinal rows or apparently distributed at random. Sensilla on the tail may be longer or stouter than those on the rest of the body, especially in the male, when they may be differentiated as **caudal** setae or spines. Specialised sensilla at the tail tip are called **terminal** setae. The region between the head and the base of the oesophagus is often called the cervical (neck) region, so that setae here may be identified as **cervical** setae.

The arrangement of the sensilla on the head is rather specialised and conforms to a characteristic basic pattern, which is generally considered to be the primitive arrangement (Fig. 2A). Six **labial** (lip) sensilla (two lateral and four submedian) surround the stoma (mouth opening) and are normally papilliform or only short setae making them difficult to detect. Posterior to the labial sensilla there is normally a circle of six (two lateral and four

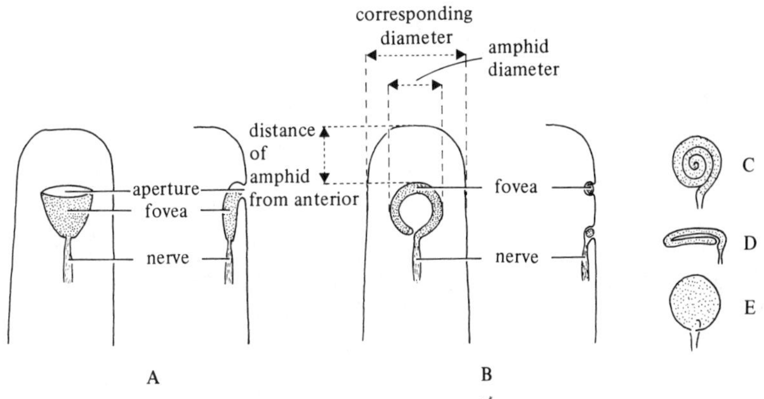

Fig. 3. Amphids. A, Pocket type. B, Spiral type with a single turn, also showing amphid measurements. C, Multi-spiralled type. D, Transverse slit type. E, Circular type.

submedian) and a circle of four (submedian) sensilla, usually setiform: anterior and posterior **cephalic** setae respectively. This basic arrangement can be denoted by the abbreviation 6 + 6 + 4 (Fig. 2B). However, the anterior cephalic sensilla may be papilliform, so that there appear to be only four setae (Fig. 2C) or the anterior and posterior circles may be situated at the same level, the latter arrangement denoted by the formula 6 + 10 (Fig. 2D). In several groups, additional setae may be found associated with the basic 10 cephalic setae. The lengths of these setae are often quoted in the literature as a proportion of the diameter of the body taken level with the posterior cephalic setae, the **head diameter** (Fig. 2D).

There are two other specialised sensilla situated laterally on the head: the **amphids**. These bilaterally symmetrical structures are usually of major importance for identification. Basically, there are two kinds, spiral or non-spiral. Usually the non-spiral amphid is a pocket-like structure (Fig. 3A) where the external opening (aperture) is in the form of a transverse slit leading to a cavity (**fovea**) filled with a gelatinous substance (**corpus gelatum**). In the spiral amphid, the fovea is open and elongated, normally turning ventrally (from the nerve) although in a few cases the amphid may be dorsally wound. When the fovea makes just one turn, the amphid may be termed loop-shaped (Fig. 3B) while if several turns are made, it is multi-spiralled (Fig. 3C). Variations on this theme include the transverse amphid (Fig. 3D) and the circular amphid (Fig. 3E) but all are essentially spiral.

In species descriptions, the amphid width is often quoted as a proportion or percentage of the corresponding body diameter (c.d.) and the distance from the anterior of the head given in terms of head diameters to the anterior amphid margin (Fig. 3B).

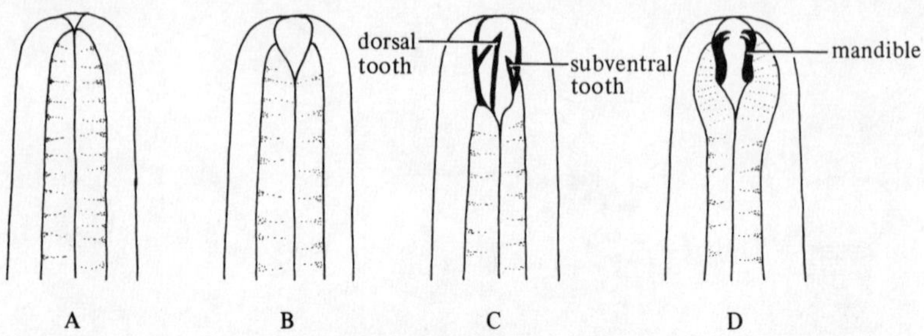

Fig. 4. Buccal cavities. A, Minute form. B, Unarmed form. C, Form with fixed teeth. D, Form with moveable mandibles.

Alimentary canal

The buccal cavity exhibits a great variety of form, reflecting the great range in feeding methods among marine nematodes. The buccal cavity may be absent or minute (Fig. 4A) to spacious but unarmed (Fig. 4B). Many species have buccal cavities armed with immoveable projections of the wall, referred to as teeth (Fig. 4C) or moveable structures termed **mandibles** or jaws (Fig. 4D). In addition, there may be rows of small **denticles** lining the buccal cavity, or there may be other projections. The detailed form of the buccal cavity is one of the most important characters for species identification.

The buccal cavity leads to the **oesophagus**, which is the muscular anterior part of the gut responsible for pumping food into the intestine. The lumen of the oesophagus is tri-radiate in cross-section (Fig. 1C). The oesophagus may be cylindrical throughout (Fig. 5A) or have a **posterior bulb** complete with a valve structure (Fig. 5B). Several nematodes have paired **pigment spots** or true **ocelli** (i.e. having a lens-like structure) situated laterally or dorsolaterally on or partly inside the anterior part of the oesophagus (Fig. 5).

About half-way along the length of the oesophagus lies the circum-oesophageal commissure or **nerve ring**, which is usually the only part of the nervous system detectable with ease. Its relative position may in some cases be of use in identification. The so-called **excretory system**, which in reality is of unproven function, consists of a single ventral gland or **rennette cell** and a duct extending anteriorly to open by a **ventral pore** somewhere in the oesophageal region (Fig. 5A). The exact postition of the pore, or its position in relation to the nerve ring or buccal cavity, is occasionally of taxonomic use. The muscular structure sometimes found at the base of the oesophagus and opening into the intestine is the oesophago-intestinal valve, usually referred to more simply as the **cardia**. The intestine itself is a straight tube formed of a single layer of cells and is taxonomically unimportant. The **rectum** connects the intestine to the **anus** in females or **cloaca** in males, and may have conspicuous muscles attached to the dorsal wall for opening the passage against the internal turgor pressure.

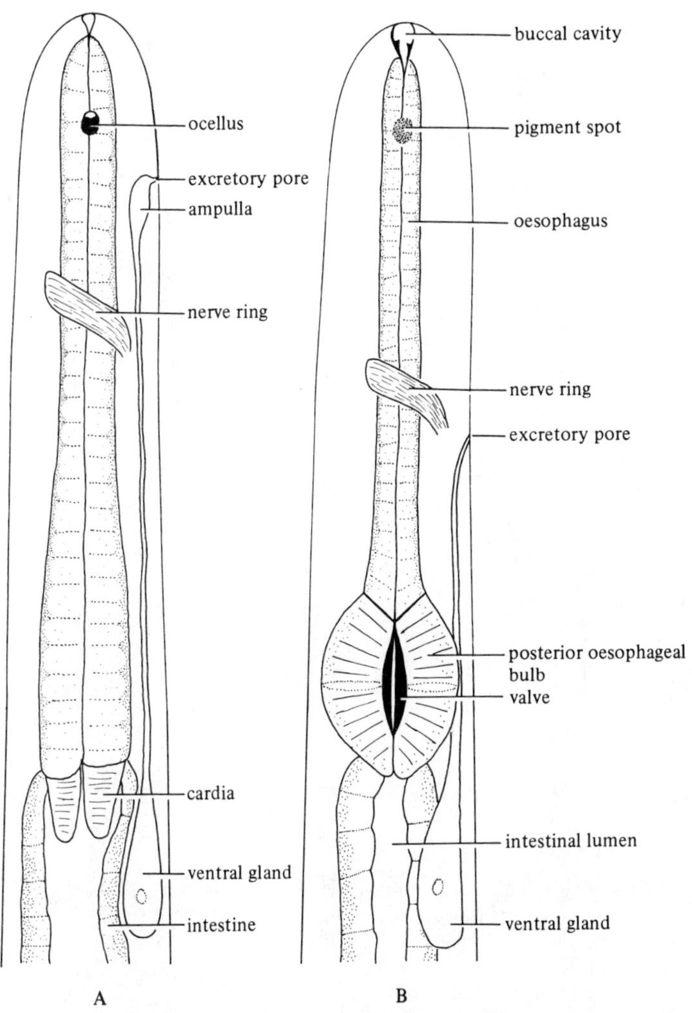

Fig. 5. Features of the anterior region. A, Showing a cylindrical oesophagus and an ocellus. B, Showing an oesophagus with a well-developed posterior bulb and a pigment spot.

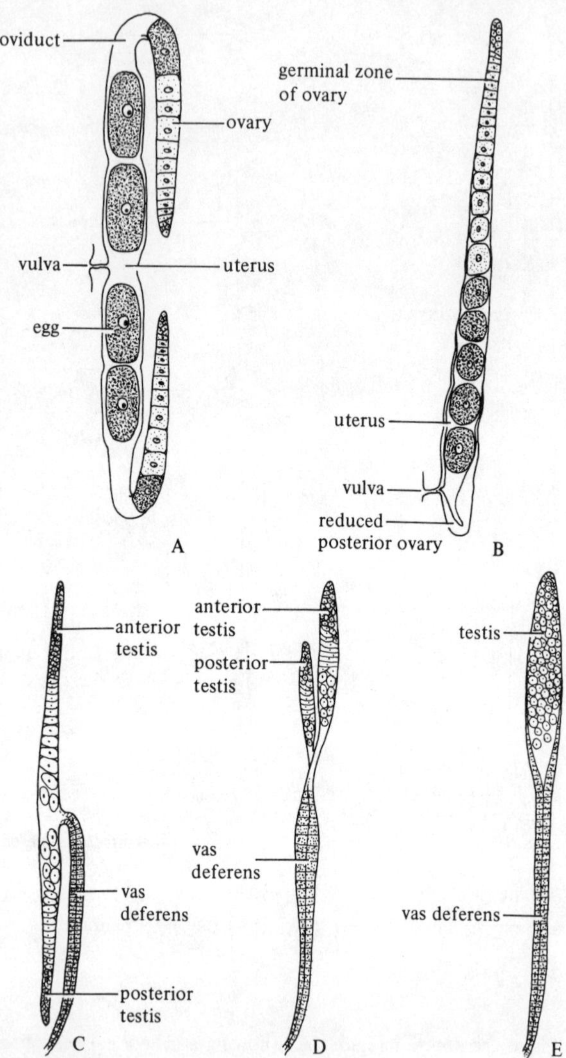

Fig. 6. Reproductive systems. A, Female with two opposed, reflexed ovaries. B, Female with single anterior outstretched ovary. C, Male with two opposed testes. D, Male with two testes arranged in tandem. E, Male with single testis.

Reproductive system

The female may have one or two **ovaries** (monodelphic/didelphic) which may or may not be reflexed (Fig. 6A, B). The number and structure of the ovaries is important and can be used to distinguish major taxonomic groups. The

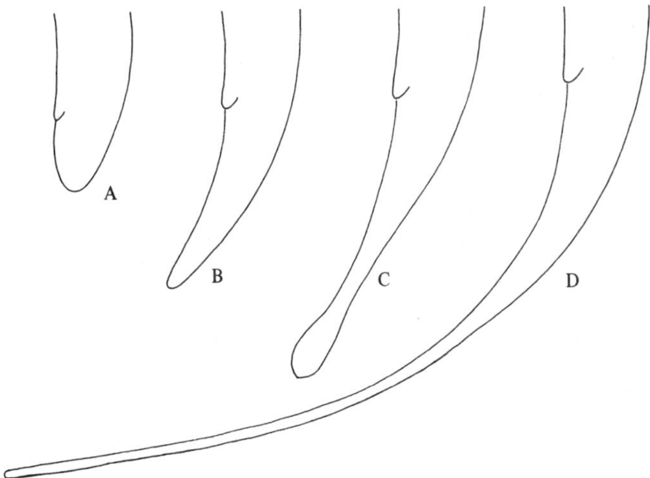

Fig. 7. Tail shapes. A, Short and round. B, Conical. C, Conico-cylindrical with swollen tip (clavate). D, Elongated, filiform.

position of the **vulva** is usually at about the middle of the body in didelphic nematodes but can be closer to the anus in monodelphic forms; its position from the anterior is given as a percentage of the total body length (V%), including the tail unless the latter is very long, when a modified percentage of anterior-to-anus length (V′%) should be given.

The male usually has two **testes** (diorchic) which may be opposed (Fig. 6C) or in tandem (Fig. 6D) but some major groups have only one (Fig. 6E). However, gonads are often difficult to distinguish in practice and are of little importance for species identification.

The most important parts of the male system for practical purposes are the cuticularised copulatory structures and accessory organs. Typically, there is a pair of cuticularised **spicules** and a guiding piece called a **gubernaculum** which lie in a sac opening into the dorsal side of the cloaca (Fig 1A). However, in practice the detailed arrangement of the **vas deferens**, rectum and spicule sac around the cloaca are difficult to elucidate and are rarely depicted. The shape and size of the spicules and gubernaculum vary considerably and are normally of great value in identification. Likewise, the presence and structure of the various kinds of **precloacal supplements** can be highly important.

Tail

The tail shape can be highly characteristic, the main themes around which there are variations being round, conical, conico-cylindrical or elongated (Fig. 7). The length of the tail is often quoted in terms of the anal (or cloacal) body diameter (a.b.d., see p. 35 and Fig. 10). The tail contains the unicellular

caudal glands, usually three (Fig. 1), which may be confined entirely to the tail or extend anterior to the anus/cloaca. The caudal glands are responsible for secreting an adhesive which exits via a specialised structure in the tail tip called a **spinneret**. Certain species may lack caudal glands entirely.

Biology

Ecology

By virtue of their wide range of adaptations, marine nematodes have exploited all seashore and seabed habitats. They are the most abundant metazoans of particulate shores, reaching their highest numbers in muddy estuaries and saltmarshes (around 20 million per square metre) and lowest in very exposed sandy shores or the deep sea sediments (0.1 million per square metre). High metabolic rate coupled with high density means that despite the generally lower biomass of the nematodes and meiofauna in general compared with the macrofauna, production values may equal or even exceed those of all the larger organisms combined.

One important feature of nematode populations is the large number of species present in any one habitat, often an order of magnitude higher than for any other major taxon (Platt and Warwick, 1980). Species richness and diversity varies among habitats, being greatest in sandy beaches with over 100 species being typical. In muddy sites and in algal communities, the number of species is more typically in the range 30–70. A more detailed community analysis shows there is variation among habitats in the degree of species dominance. In general, although seaweeds and low salinity muds have a relatively low number of species, they may be dominated by only one or a few species whereas sand and high salinity muds may have more species but none of them totally dominate the assemblage. This high species diversity can be explained by the extensive way in which nematodes are able to partition their environment, particularly as regards food.

Feeding

Potential food items include general organic detritus, decomposing organisms, bacteria, diatoms and other living organisms. Nematodes living together in the same sediment are potentially in competition for all these resources but the small size of the food gathering apparatus in relation to the food particles permits a high degree of selectivity. Indeed, the structure of the buccal cavity is often a good indicator of the feeding strategy of the individual in broad terms. Those without or with only a minute buccal cavity (Fig. 4A) are only able to ingest small particles and/or fluid and so are necessarily selective deposit feeders. Those with a buccal cavity but lacking dentition (Fig. 4B) are potentially able to ingest particles of a wider size range including diatoms and are therefore referred to conventionally as non-selective deposit feeders. However, the term 'non-selective' is probably misleading; there is now some evidence that certain nematodes may be able to selectively enhance the density of their preferred food item by a process analogous to gardening (Warwick, 1981a). Small teeth and/or denticles in the buccal cavity (Fig. 4C) enable cells to be pierced and the contents sucked out or objects scraped off surfaces; collectively these are known as epigrowth feeders. Large powerful teeth or mandibles (Fig. 4D) allow predators to seize their prey which they swallow or suck the contents. The relative proportions of each of these four feeding groups in any one community will depend on the nature and balance of the available food which in turn is often correlated with the sediment type. Much useful ecological information can be obtained simply through recognising these key groupings. One probable but as yet not fully evaluated food source is dissolved organic matter. But whatever the food, to efficiently utilise the available resources the number of offspring produced must be optimised, which is where reproductive and life-cycle strategies play an important role.

Reproduction and life cycle

Most marine nematodes are dioecious (unisexual), fertilisation is by copulation and there is direct benthic development of the egg through four juvenile stages to the adult. A few species are viviparous, the eggs hatching in the uterus. Developmental changes are associated with each of the four moults but fully formed gonads and copulatory apparatus are only found in the adults. The practical consequence of this is that juvenile stages of some species are difficult and often impossible to identify to species. Fortunately the fecundity is not so great that these juvenile stages greatly outnumber adults. The absence of a planktonic stage affects dispersal but most species are small enough to be dispersed as adults by water currents. Life cycles are typically shorter than the seasonal cycle, often only 20–30 days, and reproduction is continuous. Some of the larger nematodes may have an annual cycle. However, generation times and fecundity are markedly temperature dependent. Experimentally a 5°C increase in temperature is able to produce up to a sixfold increase in the number of eggs laid (Warwick, 1981a, b).

Phylogeny

With a virtual absence of palaeontological evidence, theories of relationships among the lower invertebrates must rely on our understanding of living forms. Morphological and embryological evidence remains equivocal but the currently accepted view is that nematodes are probably most closely related to the Nematomorpha and Gastrotricha with perhaps some more distant affinities with Rotifera and Kinorhyncha. Much of this evidence relates to the common possession of a so-called pseudocoelom. However, there is little evidence that this non-feature, the absence of a complete mesodermal lining to the body cavity, is of homologous origin.

Methodology

Collection

A great deal of useful information on how to handle marine nematodes and other meiofauna can be found in McIntyre and Warwick (in press).

Qualitative sampling by hand for intertidal or shallow subtidal sediment-dwelling nematodes is relatively easy because of their numerical abundance necessitating only small samples and the advent of SCUBA (self-contained underwater breathing apparatus) diving. For quantitative sampling, a hand-held corer of known internal diameter is by far the most suitable method and almost any type of tubing may be used, although transparent perspex is preferable since it allows a visual assessment to be made of core compression and side effects. Core compression can be minimised by the use of a tight-fitting plunger which is kept level with the sediment surface as the tube is pushed slowly in. Side effects can be minimised by avoiding the use of too narrow a tube. It is preferable to use a wide tube, internal diameter of 20 mm or more, and then sub-sample later in the laboratory if the sample would give unmanageable numbers. Many apparently anomolous results can be a consequence of a poor sampling technique right at the start of a project.

If vertical zonation of the organisms in the sediment column is of interest, a few scattered investigatory columns should be taken so as to avoid taking unnecessarily deep cores. The core should be sectioned as soon as possible after collection to avoid errors due to subsequent migration.

Remote sampling can be more of a problem because grabs and piston cores usually create a bow-wave as they descend which blows away some of the surface sediment, especially if it is light and flocculent. The surface sediment often contains a high density of meiofaunal organisms. Therefore nematode assemblages and densities reported from subsamples taken from these devices may contain serious distortions. Samplers have recently been designed which can be gently lowered onto the seabed, the process being continually monitored by television or flash photography to detect any surface disturbance. Once in place, they slowly lower a corer or battery of corers thereby simulating the actions of a SCUBA diver. Wherever possible, this is the type of device which should be used to avoid the generation of poor data at great expense. A more widespread use of deep-sea submersibles may well eventually permit the extension of the well-tested shallow water methods to the deep sea.

Quantitative sampling on rock for phytal or crevice faunas presents problems and each individual project may have to devise its own appropriate solutions. Diving is obviously the only feasible method of collecting on shallow sublittoral rock and clear plastic bags enable whole plants to be enclosed before cutting them away.

Preservation

Wherever possible, it is advisable that the animals be studied live or freshly killed because subsequent processing to glycerine often dissolves pigment spots or general body coloration. Where this is not feasible, the sample should be fixed as soon as possible in the field with cold 5% neutralised formalin. Hot (60–70°C) formalin is sometimes used to leave nematodes in a straight rather than coiled condition. In view of the dangers of inhaling hot formalin vapours this is not recommended for routine work. However, it might be of value for certain specific taxonomic studies or where previous work has shown that an unacceptable number of specimens are being fixed in a coiled condition. The addition of a small amount of the vital stain 'Rose Bengal' to the formalin can help in sorting the nematodes from detritus later. Samples can be stored in formalin indefinitely without harming the nematodes provided sufficient formalin has been added to completely fix all the organic matter. As a rough guide, there should be about seven parts of 5% formalin (in seawater) to one part of sediment by volume.

Extraction

Several methods of separating the nematodes from the sediment or phytal material have been devised and the choice of method depends largely on the type of sample (McIntyre and Warwick, in press).

For coarse sand sediments, **decantation** (Fig. 8A) is the simplest method. Sediment is placed in a measuring cylinder so that it occupies about a tenth of the volume and shaken with filtered seawater. The cylinder is allowed to stand for a few moments so that the suspended sand grains drop to the bottom leaving the nematodes in the supernatant which is then poured through a 45 μm sieve. The contents of the sieve can then be washed into a petri dish and the nematodes picked out under a dissecting microscope, their ease of detection being enhanced if Rose Bengal was added earlier. The decantation should be repeated enough times to ensure a thorough extraction. Initially keeping a record of the number extracted with each subsequent decantation will give an estimate of the efficiency. When publishing quantitative results, an estimate of extraction efficiency should always be quoted, including an assessment of any nematodes remaining in the sediment (by inspection of an aliquot). Even for a qualitative study, a complete extraction is preferable since species vary in their ease of extraction.

A more automatic extraction method for sand is **elutriation** (Fig. 8B) where the sediment is put into a glass separating funnel and filtered water (freshwater nematodes can be and often are present in tap water) run in at

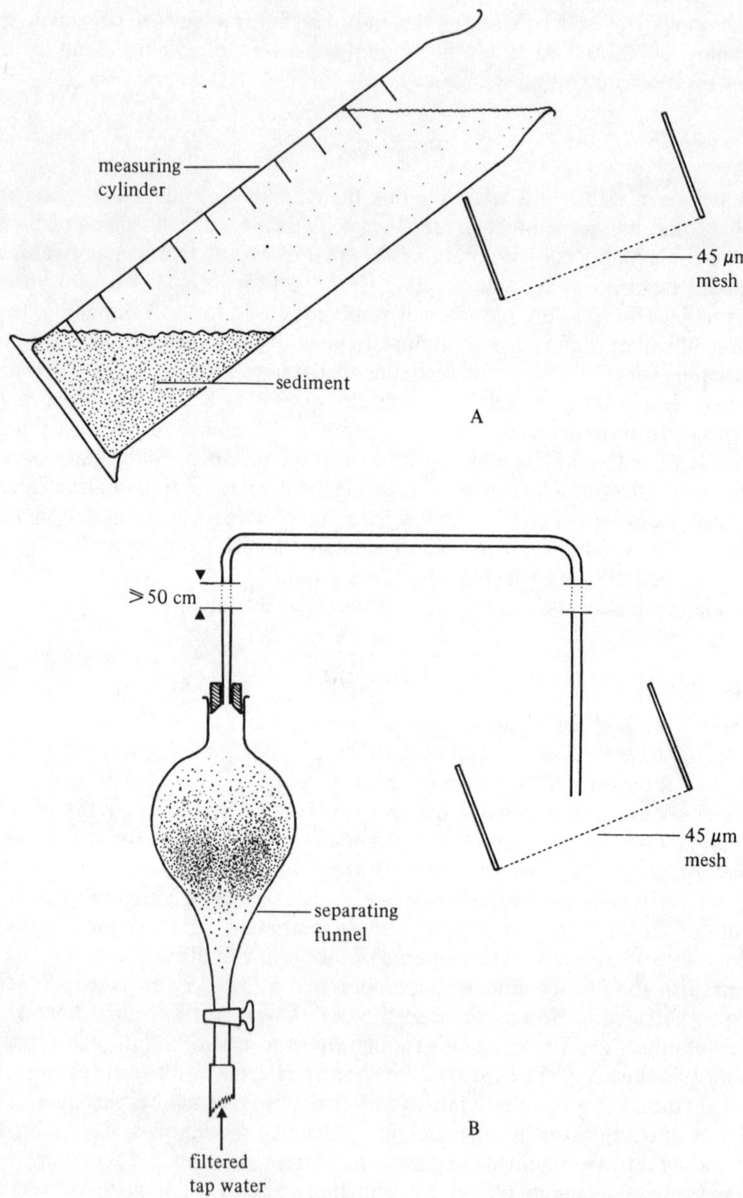

Fig. 8. Extraction by A, Decantation or B, Elutriation (see text for details).

such a rate that the sand is suspended in the flask but does not pass over onto the sieve. The system is run until a previous trial has indicated that all the nematodes have come over, which will usually be within the first 15 minutes or so. However, the sieve should be watched to make sure it has not become clogged or the sample may be lost. Again, an assessment of extraction efficiency should be quoted in published results.

Fine sands and muds are more of a problem, especially if there is a large amount of detrital material. Fine grains and detritus will not settle any faster than the nematodes and so will come over in elutriation and clog the sieve. However, a very useful **centrifugation** technique has been devised based on the liquid known as 'Ludox-TM'. Any large particles (pieces of shell etc.) are first removed and then the fine silt fraction separated from the rest of the sediment by decantation. The light fraction is then washed from the meiofauna sieve into centrifuge containers (50 ml tubes for small samples, 250 ml pots for larger ones) with tap water. A quantity of Kaolin powder (a heaped teaspoon in 250 ml pots) is added, the container shaken vigorously and spun for 7 minutes at 6000 r.p.m. The water can then be poured off easily as the Kaolin, which settles last, forms a plug over the sediment to prevent it from being decanted. The Ludox-TM, made up to a specific gravity of 1.15, is then added until the pots are nearly full and the sediment resuspended by vigorous agitation. Stirring with a spatula may be necessary to break up the Kaolin plug. The pots are then respun for 7 minutes at 6000 r.p.m. The supernatant is then poured off through a meiofauna sieve and washed thoroughly with filtered tap water before washing into a sorting dish. The extraction should be repeated twice more which should result in a separation of all but a few animals virtually free of detritus and sediment. Ludox-TM itself, which is a colloidal silica polymer, is toxic to all living organisms so that the method is only suitable for preserved material. Glassware can be cleaned by washing in dilute sodium hydroxide. It is a wise precaution to wear rubber gloves when handling Ludox-TM and sodium hydroxide.

With seaweed samples, simply shaking the plant in a container of filtered water and sieving it, together with the remains in the container in which it was stored, will provide an adequate sample. Holdfasts will have to be carefully cut apart to extract the organisms they contain.

If the nematodes are to be examined live, narcotization with 10% magnesium chloride or 1% ethanol solutions before extraction may help. Returning them to fresh seawater should restore their activity.

Mounting

For identification, the live or fixed worms can be transferred to a microscope slide in seawater, taking care to support the coverslip either with lens tissue or with appropriate sized glass rods or beads (Ballotini Manufacturing Co. Ltd., Barnsley, Yorkshire). However, this technique will not allow extended examination and when this is required the fixed animals should be processed

SPECIES DESCRIPTION: Numerical Data

Species:

Specimen no.: Sex						
Slide number						
Location						
Length (mm)						
a						
b						
c						
Head dia.						
Labial setae						
Cephalic setae						
Subcephalic setae						
Buccal cav.: 1-dia.						
Amphid : dist. from ant.						
: length – width						
: c.b.d.						
Excretory pore: from ant.						
: c.b.d.						
Nerve ring: from ant.						
: c.b.d.						
Oesophagus: length						
: c.b.d.						
: bulb						
Max. dia.						
Vulva: from ant.						
: V%						
: c.b.d.						
Gonad						
Spicule length						
a.b.d.						
Gubernaculum						
Supplements						
Tail length						
Terminal setae						

SPECIES DESCRIPTION: Non-numerical Data
Species:
Body size & shape: Cuticle: Labial setae: Cephalic setae: Somatic setae: Buccal cavity: Amphids: Oesophagus: Nerve ring: Excretory gland: Tail: Caudal glands: Spicules: Gubernaculum: Supplements: Gonad: male: Gonad: female: Eggs: Remarks:

Fig. 9. Examples of proforma lists used when recording numerical and non-numerical taxonomic data. Abbreviations: a, b, c = De Man ratios; buccal cav.: 1-dia. = length and diameter of buccal cavity; dist. = distance; ant. = anterior; c.b.d. = corresponding body diameter; Max. dia. = maximum diameter; V% = position of vulva from anterior as percentage of total body length.

to anhydrous glycerine. This can be simply achieved by transferring the nematodes to a cavity block containing by volume 5% glycerine: 5% pure ethanol: 90% freshwater, almost covering it with a coverglass and leaving it in a desiccator for a few days. This allows the ethanol and water to evaporate slowly leaving the nematodes in pure glycerine. The presence of ethanol will also take out much of the Rose Bengal if this has been used. Finally, the nematodes can be transferred to a fresh drop of anhydrous glycerine on a slide and a coverglass added, supported by glass beads or rods of an appropriate diameter. The coverglass can, if the slide is to be kept permanently, be sealed with 'Glyceel' or the two-stage process using 'Clearseal' and 'Bioseal' (Northern Biological Supplies, Norwich, Norfolk). To avoid the possibility of fungal growth in the glycerine, a few crystals of phenol can be added (WARNING: phenol is a carcinogen).

Whole mounts are usually completely sufficient for species identification but sometimes examination in apical view serves to clarify the arrangement of certain head structures such as the sensilla and teeth. An 'en face' preparation is achieved by decapitation and mounting in glycerine or warm glycerine jelly which, when cool, fixes the orientation. It will be necessary to manipulate the coverglass to obtain an exact head-on view.

Examination

Examination and identification requires a reasonably good quality microscope equipped with a ×100 oil immersion lens. Without this commitment to microscopy, frustration is guaranteed. The availability of an interference-contrast microscope is also an advantage although phase-contrast is of little use. A drawing-tube or camera-lucida will be found of considerable help.

When attempting to identify specimens, it is often more convenient to work from a drawing. Each species encountered should be drawn fairly accurately; the various features are usually in different planes of focus and it is only in the drawing that their relative positions and sizes can be visualised. The aim should be to have at least one good drawing of a male head, tail and copulatory apparatus. If the first specimen of a species encountered is a juvenile or a female, then this should be sketched with a view to replacing it subsequently if and when a male turns up. Pencil drawings of a standard needed to help identify a species from a taxonomic description in the primary literature should not take long and will provide an accurate and valuable record.

Measurements from drawings or simple outlines can be taken using either dividers for straight lines or a good quality map-measurer for curves. Appropriate calibration from a stage micrometer is then applied. Alternatively, an eyepiece micrometer can be used directly for straight measurements. Solid curved structures such as the spicules are usually measured as the chord rather than the arc. The typical measurements taken are shown in Fig. 9: proforma lists such as these can prove very useful. In descriptions, an

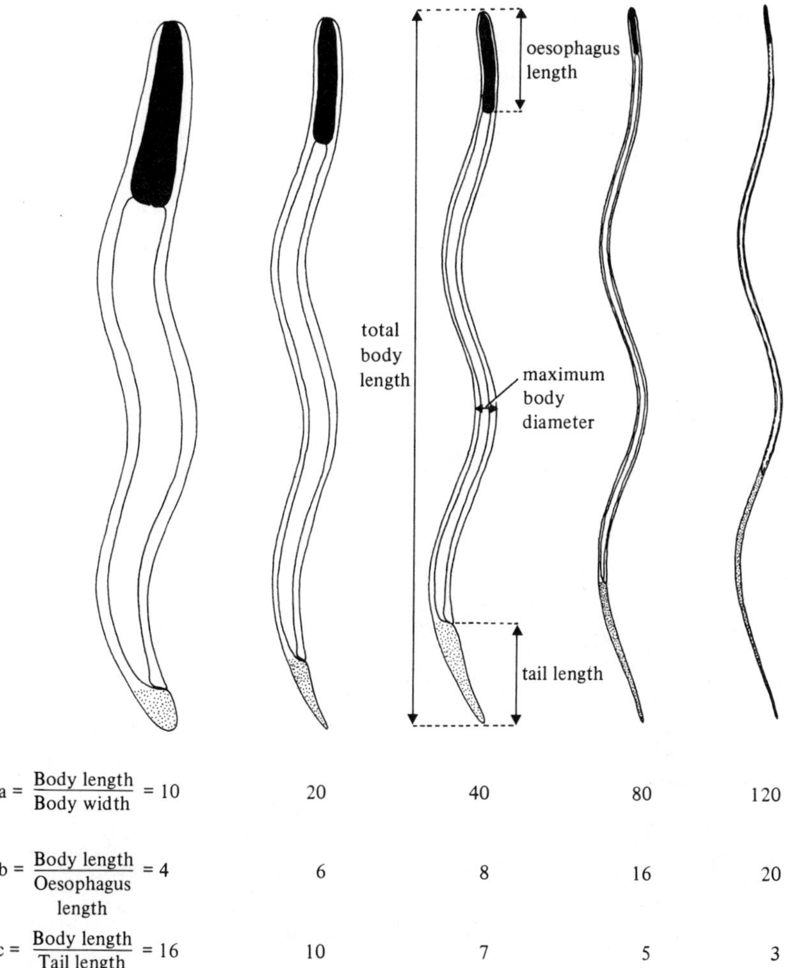

Fig. 10. Examples of De Man ratios a, b and c (see text).

indication of the relative diameter of the body, oesophagus length and tail length are often given by the De Man ratios 'a', 'b' and 'c', which are the ratios of the total body length to maximum body diameter (usually at the middle of the body), oesophagus length and tail length respectively. Fig. 10 gives examples of the appearance of nematodes when these ratios are varied.

After the investigation is completed and if the results are to be published, it is strongly recommended that voucher specimens of all the putative species are deposited in an appropriate institution with good curation facilities; in the

United Kingdom this would be the British Museum (Natural History) in London, which will store the material safely and issue the collection with a registration number. There can be no blame attached to an honest misidentification but good scientific practice would be to provide the material evidence for such a possibility to be rectified.

Classification

The Phylum Nematoda consists of two classes, the Secernentea and the Adenophorea. Only two species from the Secernentea, both members of the genus *Rhabditis*, have been found as freeliving organisms in the marine environment. For convenience, these species are dealt with in the volume on the enoplids. The rest of the freeliving marine nematodes are classified as adenophoreans. A useful review of marine nematode systematics can be found in Heip *et al.*, 1982.

The following classification of the Adenophorea follows that of Lorenzen (1981a), with minor modifications. An outline of the higher taxa is shown in Fig. 11, which excludes the Order Dorylaimida (terrestrial and freshwater forms) in addition to those families not represented in the pictorial key. The genera included in the key are listed below; it represents about 60% of the genera considered valid by Lorenzen (1981a). The omitted genera are either freshwater forms, very rare or judged to be too poorly known ever to be recognised again. Those marked with an asterisk, *, have been recorded from British waters and constitute some 55% of those in the list or 33% of those in Lorenzen (1981a).

Details of all those genera and species known up to 1973 can be found in the invaluable 'Bremerhaven Checklist of Aquatic Nematodes' compiled by Gerlach and Riemann (1973/1974).

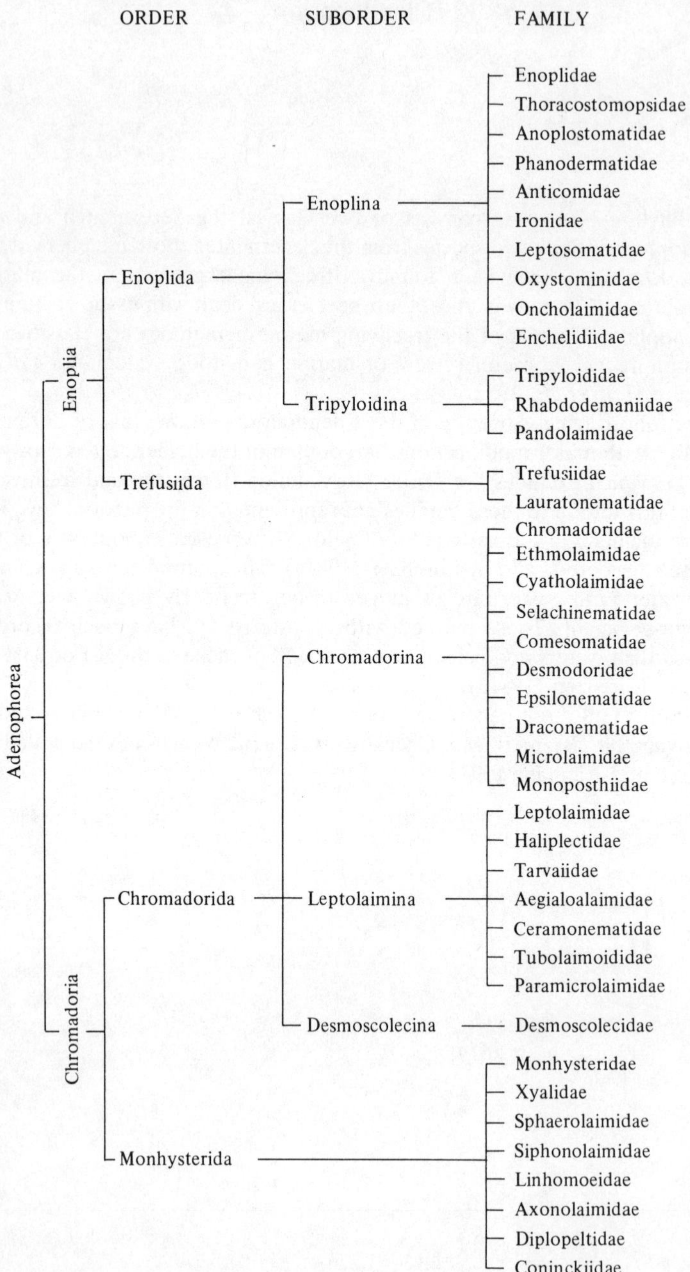

Fig. 11. Classification of the Adenophorea according to Lorenzen (1981a) excluding the Order Dorylaimida and families not represented in the pictorial key.

Class ADENOPHOREA
Subclass ENOPLIA
Order ENOPLIDA
Suborder ENOPLINA

Family ENOPLIDAE
**Enoplus* Dujardin, 1845

Family THORACOSTOMOPSIDAE
**Enoploides* Ssaweljev, 1912
**Enoplolaimus* De Man, 1893
**Epacanthion* Wieser, 1953
**Mesacanthion* Filipjev, 1927
Mesacanthoides Wieser, 1953
Oxyonchus Filipjev, 1927
**Paramesacanthion* Wieser, 1953
**Thoracostomopsis* Ditlevsen, 1918
**Trileptium* Cobb, 1933

Family ANOPLOSTOMATIDAE
**Anoplostoma* Bütschli, 1874
**Chaetonema* Filipjev, 1927

Family PHANODERMATIDAE
**Crenopharynx* Filipjev, 1934
Micoletzkyia Ditlevsen, 1926
**Phanoderma* Bastian, 1865
Phanodermella Kreis, 1928
Phanodermopsis Ditlevsen, 1926

Family ANTICOMIDAE
**Anticoma* Bastian, 1865
Anticomopsis Micoletzky, 1930
Cephalanticoma Platonova, 1976
Odontanticoma Platonova, 1976
Paranticoma Micoletzky, 1930

Family IRONIDAE

**Dolicholaimus* De Man, 1888
Parironus, Micoletzky, 1930
Pheronus Inglis, 1966
Syringolaimus De Man, 1888
**Thalassironus* De Man, 1889
**Trissonchulus* Cobb, 1920

Family LEPTOSOMATIDAE

**Cylicolaimus* De Man, 1889
Deontostoma Filipjev, 1916
**Leptosomatides* Filipjev, 1918
**Leptosomatum* Bastian, 1865
Metacylicolaimus Stekhoven, 1946
**Platycoma* Cobb, 1894
Platycomopsis Ditlevsen, 1926
**Pseudocella* Filipjev, 1927
**Synonchus* Cobb, 1894
**Thoracostoma* Marion, 1870

Family OXYSTOMINIDAE

**Halalaimus* De Man, 1888
Litinium Cobb, 1920
**Nemanema* Cobb, 1920
**Oxystomina* Filipjev, 1921
**Paroxystomina* Micoletzky, 1924
**Thalassoalaimus* De Man, 1893
Wieseria Gerlach, 1956

Family ONCHOLAIMIDAE

**Adoncholaimus* Filipjev, 1918
Filoncholaimus Filipjev, 1927
**Metaparoncholaimus* Filipjev, 1918
**Metoncholaimus* Filipjev, 1918
Meyersia Hopper, 1967
**Oncholaimellus* De Man, 1890
**Oncholaimus* Dujardin, 1845
**Pontonema* Leidy, 1855
Prooncholaimus Micoletzky, 1924
**Viscosia* De Man, 1890

Family ENCHELIDIIDAE

Bathyeurystomina Lambshead and Platt, 1979
**Belbolla* Andrássy, 1973
**Calyptronema* Marion, 1870
Ditlevsenella Filipjev, 1927
**Eurystomina* Filipjev, 1921
**Pareurystomina* Micoletzky, 1930
Polygastrophora De Man, 1922
**Symplocostoma* Bastian, 1865

Suborder TRIPYLOIDINA

Family TRIPYLOIDIDAE

**Bathylaimus* Cobb, 1894
**Gairleanema* Warwick and Platt, 1973
**Tripyloides* De Man, 1886

Family RHABDODEMANIIDAE

**Rhabdodemania* Baylis and Daubney, 1926

Family PANDOLAIMIDAE

Pandolaimus Allgen, 1929

Order TREFUSIIDA

Family TREFUSIIDAE

Cytolaimium Cobb, 1920
Halanonchus Cobb, 1920
**Rhabdocoma* Cobb, 1920
**Trefusia* De Man, 1893
Trefusialaimus Riemann, 1974

Family LAURATONEMATIDAE

Lauratonema Gerlach, 1953

Subclass CHROMADORIA
Order CHROMADORIDA
Suborder CHROMADORINA

Family CHROMADORIDAE
*_Acantholaimus_ Allgen, 1933
*_Actinonema_ Cobb, 1920
*_Atrochromadora_ Wieser, 1959
*_Chromadora_ Bastian, 1865
*_Chromadorella_ Filipjev, 1918
*_Chromadorina_ Filipjev, 1918
*_Chromadorita_ Filipjev, 1922
*_Dichromadora_ Kreis, 1929
*_Euchromadora_ De Man, 1886
*_Graphenoma_ Cobb, 1898
*_Hypodontolaimus_ De Man, 1886
*_Innocuonema_ Inglis, 1969
*_Neochromadora_ Micoletzky, 1924
Nygmatonchus Cobb, 1933
Parachromadorita Blome, 1974
*_Parapinnanema_ Inglis, 1969
*_Prochromadora_ Filipjev, 1922
*_Prochromadorella_ Micoletzky, 1924
*_Ptycholaimellus_ Cobb, 1920
*_Rhips_ Cobb, 1920
*_Spiliphera_ Bastian, 1865
*_Spilophorella_ Filipjev, 1917
*_Steineridora_ Inglis, 1969
Trochamus Boucher and Bovée, 1972

Family ETHMOLAIMIDAE
Ethmolaimus De Man, 1880
*_Filitonchus_ Platt, 1982
Gomphionchus Platt, 1982
Gomphionema Wieser and Hopper, 1966
*_Nannolaimus_ Cobb, 1920
*_Neotonchoides_ Platt, 1982
*_Neotonchus_ Cobb, 1933
*_Trichethmolaimus_ Platt, 1982

Family CYATHOLAIMIDAE

Acanthonchus Cobb, 1920
**Cyatholaimus* Bastian, 1865
Kraspedonema Gerlach, 1954
Longicyatholaimus Micoletzky, 1924
**Marylynnia* Hopper, 1977
Metacyatholaimus Stekhoven, 1942
Minolaimus Vitiello, 1970
Nannolaimoides Ott, 1972
**Paracanthonchus* Micoletzky, 1924
Paracyatholaimoides Gerlach, 1953
**Paracyatholaimus* Micoletzky, 1922
**Paralongicyatholaimus* Stekhoven, 1942
**Pomponema* Cobb, 1917
**Praeacanthonchus* Micoletzky, 1924

Family SELACHINEMATIDAE

Cheironchus Cobb, 1917
**Choniolaimus* Ditlevsen, 1918
Demonema Cobb, 1894
**Gammanema* Cobb, 1920
**Halichoanolaimus* De Man, 1886
Latronema Wieser, 1954
**Richtersia* Steiner, 1916
**Synonchiella* Cobb, 1933
Synonchium Cobb, 1920

Family COMESOMATIDAE

Cervonema Wieser, 1954
Comesoma Bastian, 1865
Comesomoides Gourbault, 1980
**Dorylaimopsis* Ditlevsen, 1918
Hopperia Vitiello, 1969
**Laimella* Cobb, 1920
Metacomesoma Wieser, 1954
**Paracomesoma* Hope and Murphy, 1972
Paramesonchium Hopper, 1967
Pierrickia Vitiello, 1970
**Sabatieria* Rouville, 1903
Vasostoma Wieser, 1954

Family DESMODORIDAE
Acanthopharynx Marion, 1870
**Catanema* Cobb, 1920
**Chromaspirina* Filipjev, 1918
**Desmodora* De Man, 1889
Eubostrichus Greef, 1869
**Leptonemella* Cobb, 1920
**Metachromadora* Filipjev, 1918
Metadesmodora Stekhoven, 1942
**Molgolaimus* Ditlevsen, 1921
**Onyx* Cobb, 1891
Paradesmodora Stekhoven, 1950
Parallelocoilas Boucher, 1975
Polysigma Cobb, 1920
**Pseudonchus* Cobb, 1920
**Sigmophoranema* Hope and Murphy, 1972
**Spirinia* Gerlach, 1963

Family EPSILONEMATIDAE
**Epsilonema* Steiner, 1927
Metepsilonema Steiner, 1927
Perepsilonema Lorenzen, 1973

Family DRACONEMATIDAE
**Dracognomus* Allen and Noffsinger, 1978
**Draconema* Cobb, 1913
**Notochaetosoma* Irwin-Smith, 1918
**Paradraconema* Allen and Noffsinger, 1978
Prochaetosoma Micoletzky, 1922

Family MICROLAIMIDAE
Aponema Jensen, 1978
**Bolbolaimus* Cobb, 1920
**Calomicrolaimus* Lorenzen, 1976
Ixonema Lorenzen, 1971
**Microlaimus* De Man, 1880

Family MONOPOSTHIIDAE
**Monoposthia* De Man, 1889
Monoposthoides Hopper, 1963
**Nudora* Cobb, 1920
Rhinema Cobb, 1920

Suborder LEPTOLAIMINA

Family LEPTOLAIMIDAE

**Antomicron* Cobb, 1920
**Camacolaimus* De Man, 1889
**Cricolaimus* Southern, 1914
**Dagda* Southern, 1914
Deontolaimus De Man, 1880
**Diodontolaimus* Southern, 1914
**Halaphanolaimus* Southern, 1914
Leptolaimoides Vitiello, 1971
**Leptolaimus* De Man, 1876
**Onchium* Cobb, 1920
Procamacolaimus Gerlach, 1954
**Stephanolaimus* Ditlevsen, 1914

Family HALIPLECTIDAE

**Haliplectus* Cobb, 1913
Setoplectus Vitiello, 1971

Family TARVAIIDAE

**Tarvaia* Allgen, 1934

Family AEGIALOALAIMIDAE

Aegialoalaimus De Man, 1907
**Cyartonema* Cobb, 1920
Diplopeltoides Gerlach, 1962

Family CERAMONEMATIDAE

**Ceramonema* Cobb, 1920
**Dasynemoides* Chitwood, 1936
Metadasynemella De Coninck, 1942
Metadasynemoides Haspeslagh, 1973
**Pselionema* Cobb, 1933
**Pterygonema* Gerlach, 1954

Family TUBOLAIMOIDIDAE

**Chitwoodia* Gerlach, 1956
Tubolaimoides Gerlach, 1963

Family PARAMICROLAIMIDAE

Paramicrolaimus Wieser, 1954

Suborder DESMOSCOLECINA

Family DESMOSCOLECIDAE

Calligyrus Lorenzen, 1969
Desmogerlachia Freudenhammer, 1975
Desmolorenzenia Freudenhammer, 1975
**Desmoscolex* Claparède, 1863
**Gerlachius* Andrássy, 1976
Greeffiella Cobb, 1922
Greeffiellopsis Schrage and Gerlach, 1975
Hapalomus Lorenzen, 1969
Meylia Gerlach, 1956
Pareudesmoscolex Weischer, 1962
Quadricoma Filipjev, 1922
**Tricoma* Cobb, 1893

Order MONHYSTERIDA

Family MONHYSTERIDAE

Diplolaimella Allgen, 1929
Diplolaimelloides Meyl, 1954
Gammarinema Kinne and Gerlach, 1953
**Monhystera* Bastian, 1865

Family XYALIDAE

Ammotheristus Lorenzen, 1977
**Amphimonhystera* Allgen, 1929
Amphimonhystrella Timm, 1961
**Cobbia* De Man, 1907
**Daptonema* Cobb, 1920
Echinotheristus Thun and Riemann, 1967
Elzalia Gerlach, 1957
Gnomoxyala Lorenzen, 1977
**Gonionchus* Cobb, 1920
Linhystera Juario, 1974
**Metadesmolaimus* Stekhoven, 1935
Omicronema Cobb, 1920
**Paramonohystera* Steiner, 1916
Promonhystera Wieser, 1956
Pseudosteineria Wieser, 1956
Retrotheristus Lorenzen, 1977
Rhynchonema Cobb, 1920
Scaptrella Cobb, 1917
Steineria Micoletzky, 1922
Stylotheristus Lorenzen, 1977
**Theristus* Bastian, 1865
**Trichotheristus* Wieser, 1956
Valvaelaimus Lorenzen, 1977
Xenolaimus Cobb, 1920
**Xyala* Cobb, 1920

Family SPHAEROLAIMIDAE

Doliolaimus Lorenzen, 1966
**Parasphaerolaimus* Ditlevsen, 1918
**Sphaerolaimus* Bastian, 1865
Subsphaerolaimus Lorenzen, 1978a

Family SIPHONOLAIMIDAE

**Siphonolaimus* De Man, 1893

Family LINHOMOEIDAE

Anticyathus Cobb, 1920
***Desmolaimus* De Man, 1880
Didelta Cobb, 1920
***Disconema* Filipjev, 1918
***Eleutherolaimus* Filipjev, 1922
***Eumorpholaimus* Schulz, 1932
***Linhomoeus* Bastian, 1865
Megadesmolaimus Wieser, 1954
***Metalinhomoeus* De Man, 1907
***Paralinhomoeus* De Man, 1907
***Terschellingia* De Man, 1888

Family AXONOLAIMIDAE

Apodontium Cobb, 1920
***Ascolaimus* Ditlevsen, 1919
***Axonolaimus* De Man, 1889
***Odontophora* Bütschli, 1874
Odontophoroides Boucher and Helléouët, 1977
Parodontophora Timm, 1963
Pseudolella Cobb, 1920
Synodontium Cobb, 1920

Family DIPLOPELTIDAE

***Araeolaimus* De Man, 1888
***Campylaimus* Cobb, 1920
***Diplopeltis* Cobb in Stiles and Hassal, 1905
***Diplopeltula* Gerlach, 1950
Southerniella Allgen, 1932

Family CONINCKIIDAE

Coninckia Gerlach, 1956

Pictorial key to genera

The key is designed as a purely practical means of identifying to genus most of the specimens that are likely to be encountered. The arrangement of the figures is based on morphological similarity but does not necessarily reflect systematic relationships among the taxa. The key itself consists of 25 figures, the first four (Figs. 12–15) being attempts to short-cut the main part which begins at Fig. 16 and finishes with Fig. 36. Each figure then consists of drawings of the head and tail regions of several genera. These drawings are not of inidividual species, except for monotypic genera, but are idealised 'caricatures' designed to show the diagnostic features of each genus. Where a character is variable within a genus, the representation is biased towards the more commonly reported species. Important diagnostic characters, including some which may not be illustrated, are given in the figure legends. It should be remembered that each drawing and pen-picture in the legend is only a composite and does not constitute a strict generic definition.

In each of the Figs. 16–36, the more common genera are arranged in the left-hand column, so by scanning down only those on the left it is likely that one will be found which resembles the specimen in question. If the fit is still not satisfactory, those genera further to the right should be checked. In general, the further to the right a genus is, the less frequently it has been reported and also tends to be less well known. It should be noted that not all genera are figured; if the head and tail pictures would be identical to another genus it is omitted, but is included in the figure legend. Also in the figure legends, characters mentioned for the genus on the left-hand side of the page generally apply also to the genera to the right unless specifically contradicted. Genera with an asterisk* have been recorded from the British Isles. Those with a double asterisk** have species described in this volume on the page indicated.

The following abbreviations are used: a.b.d. = anal or cloacal body diameter(s); c.d. or c.b.d. = corresponding (body) diameter(s); h.d. = head diameter(s); L = length in millimetres.

Fig. 12. Desmoscolecidae. Most marine nematodes have a fusiform shape (see Fig. 1). However, desmoscolids are characterised by short fat bodies, usually with conspicuous transverse rings (desmen) or rows of spines.

A, *Desmoscolex**, 12–44 oval desmen, most with 17; L = 0.1–0.5.
B, *Desmolorenzenia*, 18 asymmetrical desmen with a double desmen where asymmetry changes direction; L = 0.2–0.3.
C, *Pareudesmoscolex*, desmen absent; cuticle with warts; 65–100 cuticular rings; L = 0.1–0.3.
D, *Greeffiella*, transverse rows of setae and spines; L = 0.2–0.4.
E, *Calligyrus*, no setae but scale-like spines especially developed on tail; L = 0.2–0.3.
F, *Greeffiellopsis*, similar to *Greeffiella* except: rows of small spines between annules; four long setae on tail tip; L = 0.2.
G, *Hapalomus*, fringe of tiny hairs seen only in optical section; L = 0.1–0.2.
H, *Tricoma**, 29–240 oval desmen, most with 60–80; L = 0.2–1.3.
I, *Quadricoma*, 33–66 asymmetrical desmen which change direction at some point; L = 0.2–1.4.
J, *Desmogerlachia*, desmen absent; cuticle with warts; 190–270 cuticular rings; L = 0.4–1.2.

Note: Compare *Meylia* (Fig. 29B), *Gerlachius* (Fig. 29A) and *Richtersia* (Fig. 34H), all of which have short fat bodies also.

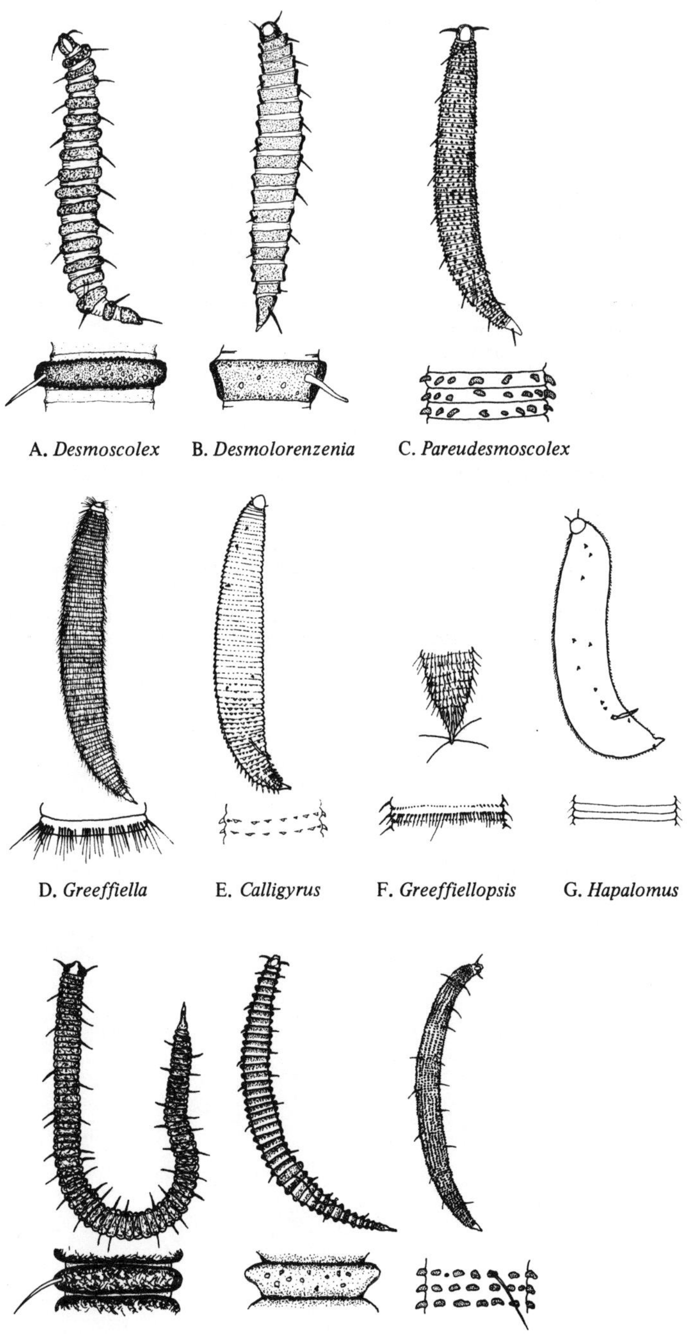
A. *Desmoscolex*
B. *Desmolorenzenia*
C. *Pareudesmoscolex*
D. *Greeffiella*
E. *Calligyrus*
F. *Greeffiellopsis*
G. *Hapalomus*
H. *Tricoma*
I. *Quadricoma*
J. *Desmogerlachia*

Fig. 13. Draconematidae and Epsilonematidae. Two groups have a characteristic 'S' or 'ε' shaped body provided with rows of stout ventral spines used for locomotion (ambulatory setae) and often also have a swollen anterior region.

A, ***Draconema****, 'S'-shaped body; ventral ambulatory setae in posterior third of body; conspicuous U-shaped amphids on the non-annulated part of head (rostrum); bulbous oesophageal region; oesophagus with median isthmus; first 7–14 body annules posterior to rostrum larger than succeeding ones; L = 0.4–1.7.

*Paradraconema**, similar to *Draconema* except: prominently enlarged post-rostral annules absent; L = 0.6–1.5.

B, C, *Prochaetosoma*, similar to *Draconema* except: oesophageal region not so bulbous; oesophagus lacks isthmus; buccal cavity with dorsal tooth; oesophagus bulb with cuticularised valve; L = 0.5–1.0.

D, *Dracognomus**, similar to *Prochaetosoma* except: amphid inconspicuous and situated at base of rostrum; L = 0.3–0.6.

E, *Notochaetosoma**, similar to *Prochaetosoma* except: dorsal tooth absent; rostral cuticle very thick; L = 0.9–1.6.

F, G, ***Epsilonema****, 'ε'-shaped body; four rows of ventral ambulatory setae in middle of body; four cephalic setae; 6–8 subcephalic setae; L = 0.3–0.5.

H, *Metepsilonema*, similar to *Epsilonema* except: only two subcephalic setae; L = 0.3–0.5.

I, *Perepsilonema*, similar to *Epsilonema* except: ambulatory setae absent; of the 6–8 subcephalic setae, four situated immediately posterior to cephalic setae; L = 0.3–0.5.

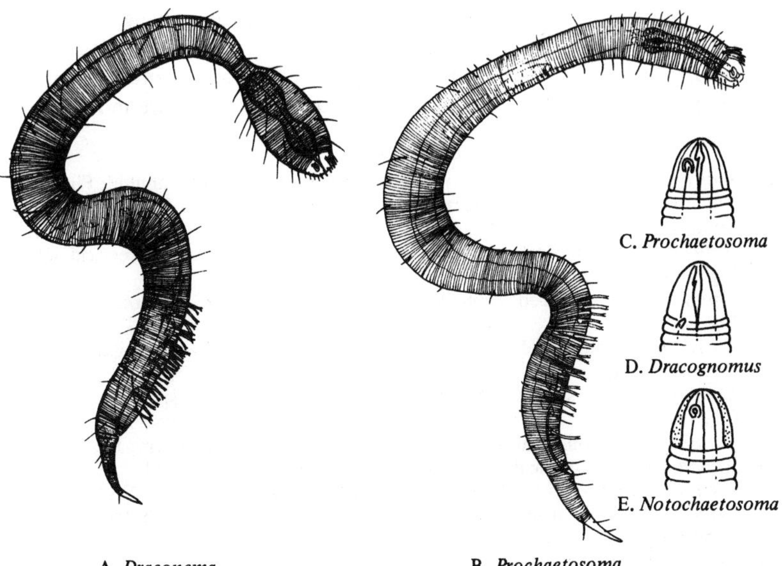

A. *Draconema*

B. *Prochaetosoma*

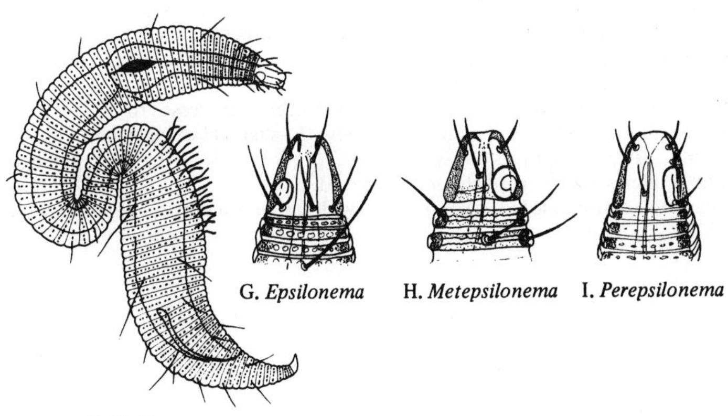

F. *Epsilonema*

Fig. 14. Star groups. If the body is the normal fusiform shape, it may be possible to short-cut the key by comparing the specimen with the following pictures, which represent the 21 most frequently recorded British genera. If the specimen resembles one of these, go direct to the figure indicated.

A, *Bathylaimus*, big buccal cavity; jointed cephalic setae; circular amphid; Fig. 16A.
B, *Thalassoalaimus*, reduced buccal cavity; pocket amphids; 6 + 4 or 6 + 6 cephalic setae; Fig. 16G.
C, *Anticoma*, small buccal cavity; 10 cephalic setae; row of lateral cervical setae; Fig. 18A.
D, *Thoracostoma*, reduced cephalic setae; cephalic capsule present; short tail; Fig. 17E.
E, *Enoplolaimus*, three hollow mandibles; lips extended; labial setae well developed; Fig. 19E.
F, *Enoploides*, three solid mandibles; labial setae well developed; spicules elongated; Fig. 19J.
G, *Enoplus*, three solid mandibles; lips low; labial sensilla papilliform; Fig. 19I.
H, *Anoplostoma*, cylindrical buccal cavity; copulatory bursa; Fig. 18I.
I, *Viscosia*, elongated buccal cavity; three teeth; Fig. 20A.
J, *Leptolaimus*, circular amphid; cuticle conspicuously annulated; tubular precloacal supplements; Fig. 22A.
K, *Camacolaimus*, stylet in buccal cavity; Fig. 22I.
L, *Axonolaimus*, conical buccal cavity, loop-shaped amphid; Fig. 23A.
M, *Metalinhomoeus*, small buccal cavity; circular amphid; four cephalic setae; Fig. 25A.
N, *Daptonema*, 10 cephalic setae; circular amphid; Fig. 26I.
O, *Monoposthia*, cuticle with longitudinal 'V'-shaped markings; four cephalic setae; Fig. 28E.
P, *Spirinia*, circular loop amphid surrounded by striations; Fig. 29J.
Q, *Microlaimus*, circular amphid posterior to head; four cephalic setae; Fig. 29C.
R, *Neochromadora*, conical buccal cavity with small teeth; cuticle ornamented with transverse rows of dots; Fig. 35E.
S, *Euchromadora*, conical buccal cavity with teeth; transverse rows of cuticle ornamentation; copulatory apparatus with extra 'L'-shaped lateral pieces; Fig. 36I.
T, *Sabatieria*, spiral amphid; transverse rows of dots which may be arranged randomly laterally; four cephalic setae; Fig. 31A.
U, *Pomponema*, spiral amphid; buccal cavity with pointed teeth; six cephalic setae; Fig. 32K

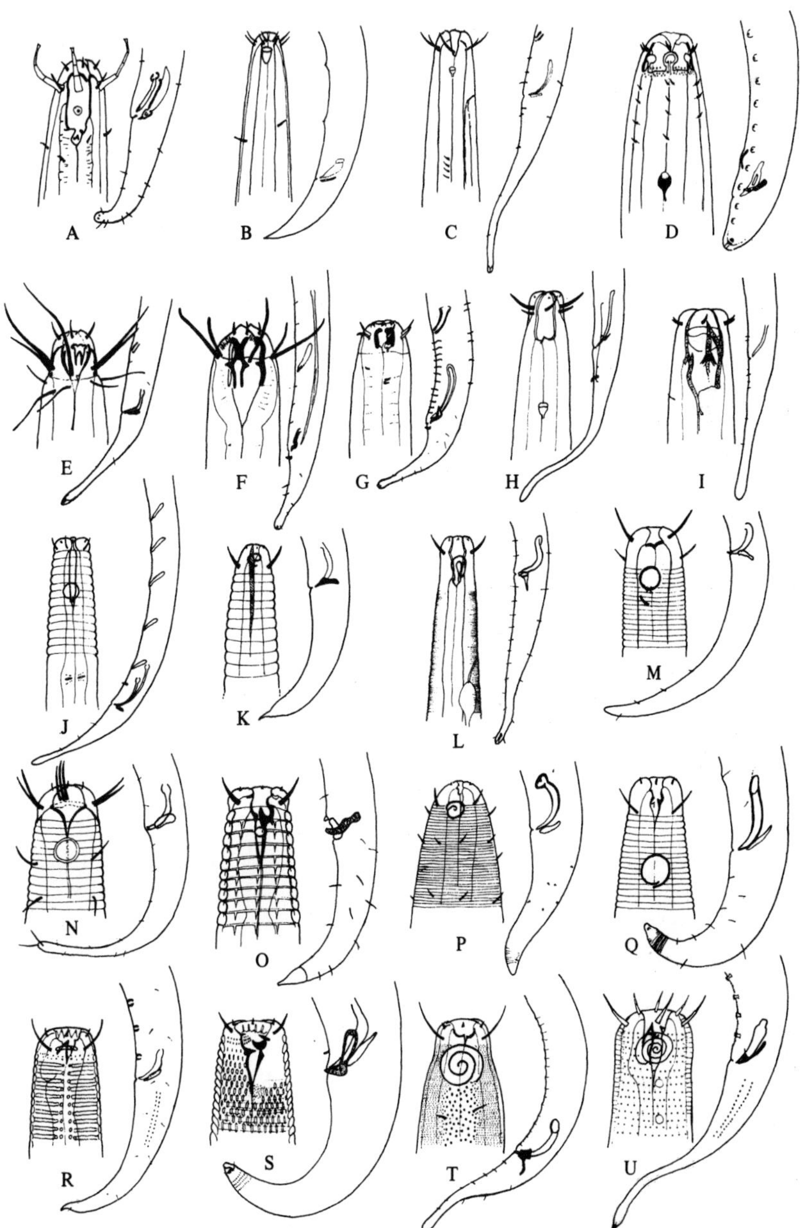

A
B
C
D
E
F
G
H
I
J
K
L
M
N
O
P
Q
R
S
T
U

Fig. 15. Cuticle patterns. If the specimen does not appear to be one of the star groups, it may still be possible to get into the key by looking at the cuticle pattern, although this is by no means an infallible guide since many exceptions exist within groups. However, observe the cuticle in the posterior oesophageal region. If it conforms to any of the following statements, turn direct to the figure indicated.

A, Cuticle smooth without dots; Fig. 16.
B, Cuticle coarsely striated resembling annulations; Fig. 22.
C, Cuticle externally fairly smooth but striations appear to be the result of a deeper structure; Fig. 23.
D, Cuticle marked with transverse rows of dots; Fig. 31.
E, Cuticle covered with fine dots, sometimes irregular laterally, although the surface appears smooth; Fig. 31.
F, Cuticle marked with longitudinal rows of structures; Fig. 28.

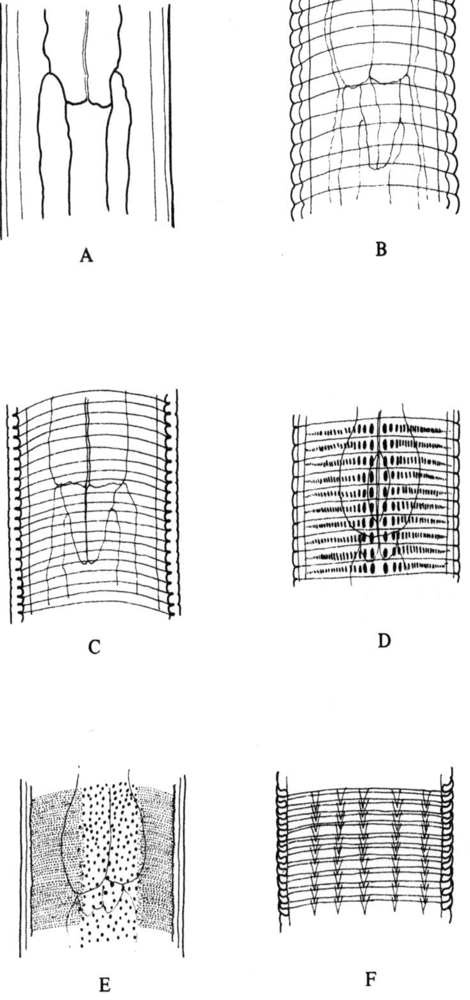
A
B
C
D
E
F

Fig. 16.

A, ***Bathylaimus*****, jointed cephalic setae; amphid a rounded spiral; buccal cavity in two sections; large gubernaculum; L = 1.5–2; p. 264.

B, *Tripyloides***, buccal cavity in several sections; amphid circular; L = 1–2; p. 278.

C, *Gairleanema***, large labial setae with bulbous swellings; dorsal tooth; precloacal supplements; L = 2; p. 276.

D, ***Trefusia*****, buccal cavity absent; amphids pocket-like; cephalic setae jointed; cervical papillae present; tail long or short; female didelphic; L = 2–3; p. 290.

E, *Rhabdocoma***, circular amphid; cervical papillae present; 12–16 precloacal papillae; female monodelphic with ovary posterior to vulva; L = 2–3; p. 288. *Cytolaimium*, similar to *Rhabdocoma* except: cervical papillae absent; 7–8 post-cloacal papillae; female didelphic; L = 3.

F, *Trefusialaimus*, 10 short cephalic setae; pericloacal papillae; single testis; L = 7–8.

G, ***Thalassoalaimus*****, buccal cavity absent; 10–12 cephalic setae; four subcephalic setae; amphid pocket-like; precloacal papillae usually; L = 1–3; p. 202.

H, *Litinium*, male amphid horse-shoe shaped (inset, female amphid); L = 1–2.

I, *Wieseria*, amphid level with or posterior to subcephalic setae; precloacal spine; L = 1–3.

J, ***Oxystomina*****, buccal cavity absent; six cephalic setae; four subcephalic setae (may be reduced); L = 1.5–3; p. 193.

K, *Halalaimus***, amphid a longitudinal slit; L = 1–3; p. 190.

L, *Nemanema***, similar to *Oxystomina* except: tail rounded; L = 3; p. 198.

M, *Paroxystomina***, six short cephalic setae; small conical buccal cavity; cervical setae present; two subventral rows of winged precloacal supplements; female has supplements anterior and posterior to vulva; L = 2–6; p. 200.

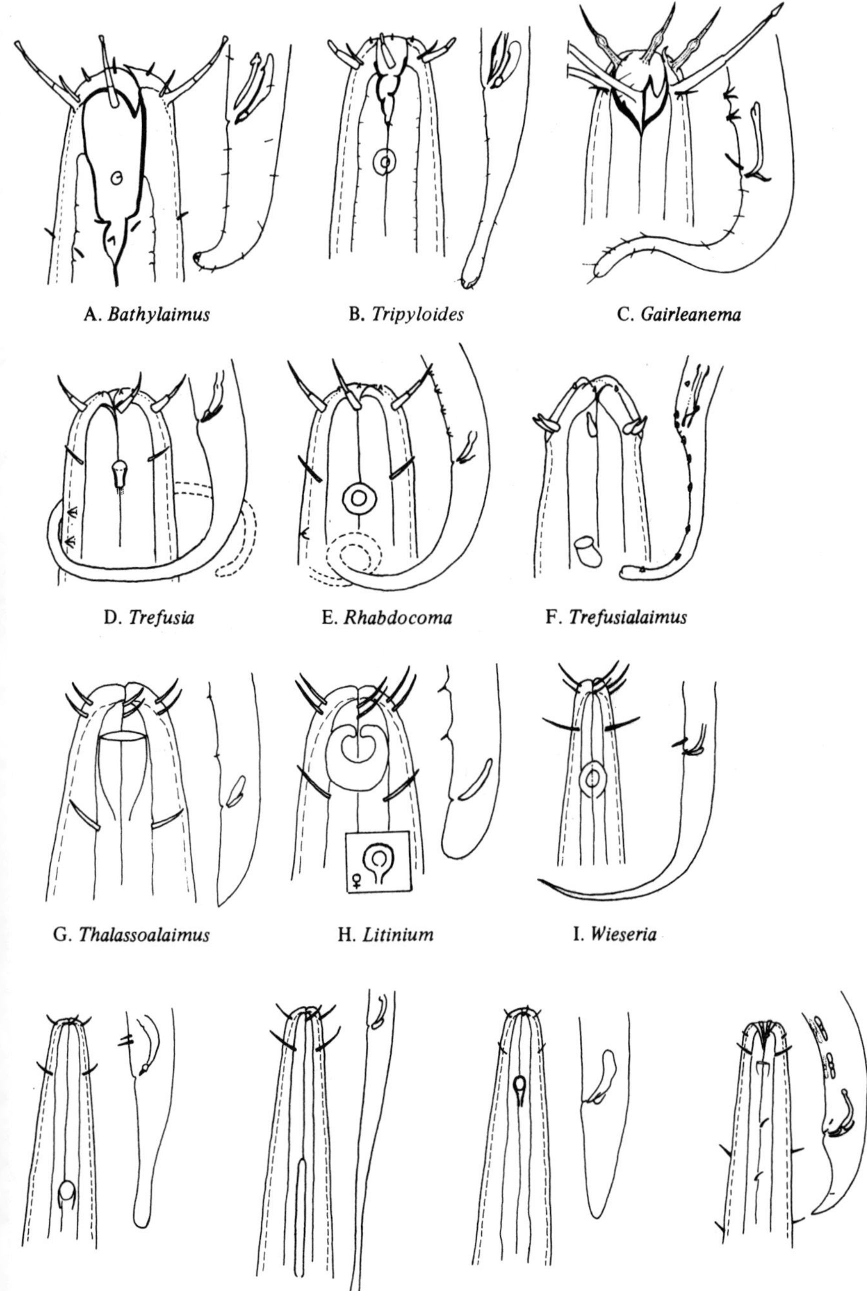
A. Bathylaimus
B. Tripyloides
C. Gairleanema
D. Trefusia
E. Rhabdocoma
F. Trefusialaimus
G. Thalassoalaimus
H. Litinium
♀
I. Wieseria
J. Oxystomina
K. Halalaimus
L. Nemanema
M. Paroxystomina

Fig. 17.

A, ***Leptosomatum*****, buccal cavity absent; cephalic setae papilliform; ocellus present; L = 8–13; p. 176.

B, *Leptosomatides***, precloacal supplement; two subventral rows of setae on papillae anterior to supplement; gubernaculum with apophysis; L = 9–18; p. 174.

C, *Platycomopsis*, cephalic sensilla setiform; ocellus absent; 2–4 long cervical setae posterior to amphid; precloacal supplement; L = 13–22.

D, *Platycoma***, three small teeth anteriorly; two stout setae associated with amphid of male only; tail pointed; L = 9–11; p. 178.

E, ***Thoracostoma*****, cephalic capsule well developed with ventral hollow tooth-like structure (tropis); ocellus present; gubernaculum apophysis almost parallel to spicules; L = 5–10; p. 188.

F, *Pseudocella***, ventral tropis present or absent; ocellus absent; gubernaculum apophysis at right angles to spicules; L = 8–12; p. 180.

G, *Deontostoma*, tropis absent; ocellus present (usually) or absent; L = 15–26.

H, ***Cylicolaimus*****, buccal cavity large; one dorsal and two subventral teeth; precloacal supplement and subventral spines; gubernaculum with apophysis; L = 23–30; p. 172.

I, *Synonchus***, two subventral onchia and dorsal tooth; precloacal supplement and two subventral rows of papillae and setae; tail clavate; L = 8–20; p. 182.

J, *Metacylicolaimus*, dorsal tooth; tail filiform; precloacal supplement (usually); L = 7–14.

K, ***Phanoderma*****, trilobed cephalic capsule; often a striated cervical capsule present; posterior oesophagus has a crenellate cellular structure; ocellus present or absent; precloacal supplement; L = 3–6; p. 156.

L, *Crenopharynx***, attenuated head; amphid indistinct; cephalic capsule weak; cervical capsule absent; buccal cavity with cuticularised structures; ocellus absent; precloacal supplements absent; spicules long; L = 4–8; p. 154.

Phanodermella, similar to *Crenopharynx* except: spicules short; precloacal supplement present; L = 3–4.

M, *Phanodermopsis*, head attenuated; buccal cavity simple; cephalic capsule weak; precloacal supplement absent; L = 4–7.

N, *Micoletzkyia*, head set-off; cephalic capsule weak; precloacal supplement present; L = 3–9.

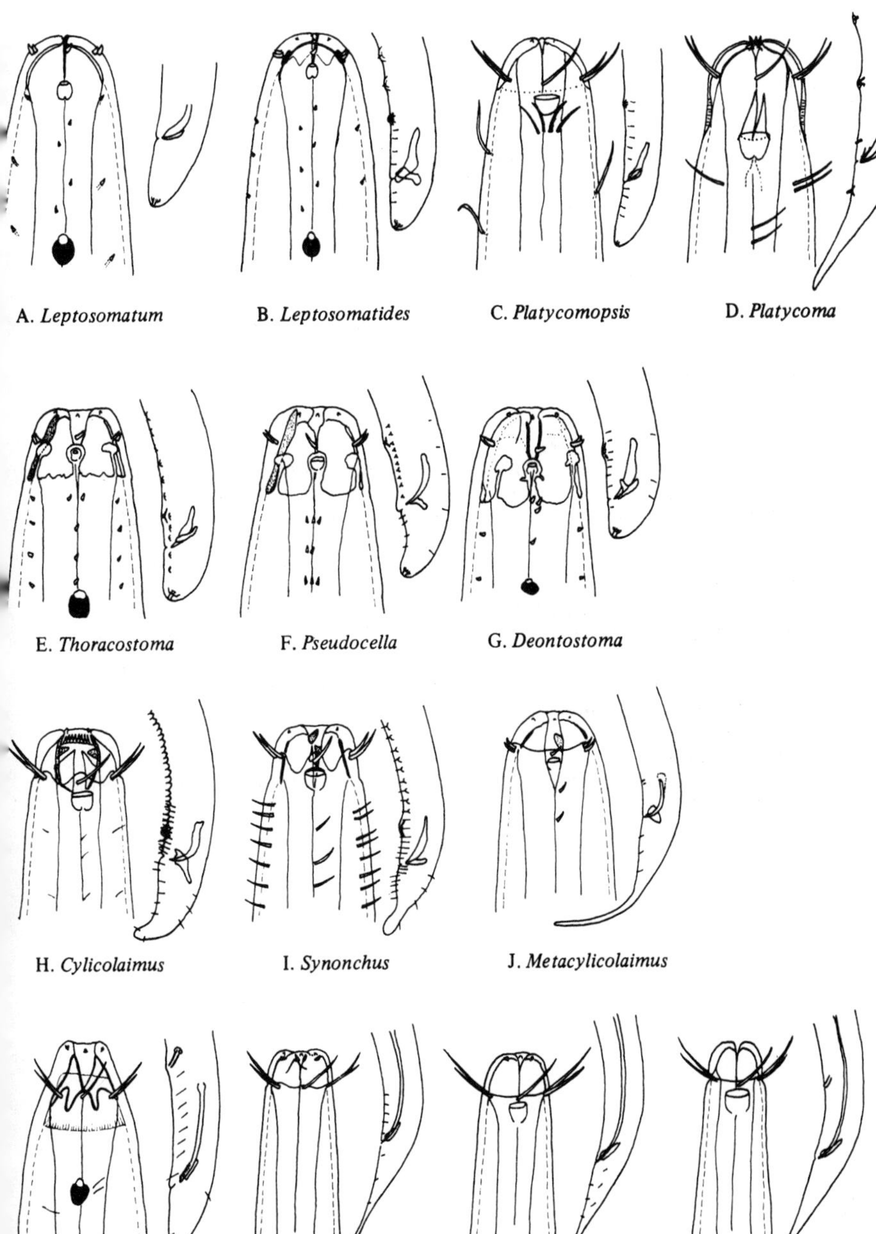
A. Leptosomatum
B. Leptosomatides
C. Platycomopsis
D. Platycoma
E. Thoracostoma
F. Pseudocella
G. Deontostoma
H. Cylicolaimus
I. Synonchus
J. Metacylicolaimus
K. Phanoderma
L. Crenopharynx
M. Phanodermopsis
N. Micoletzkyia

Fig. 18.

A, ***Anticoma*****, buccal cavity small; row of lateral cervical setae; precloacal supplement; L = 1–4; p. 160.
Anticomopsis, similar to *Anticoma* except: gubernaculum and precloacal supplement absent; L = 1–3.

B, *Paranticoma*, excretory pore on papilla; precloacal supplement absent; L = 2–5.

C, *Odontanticoma*, three teeth at anterior of oesophagus; L = 2–3.

D, *Cephalanticoma*, cephalic capsule well developed; three teeth at anterior of oesophagus; excretory pore opens posterior to cervical setae; tail long; L = 4–5.

E, ***Thalassironus*****, buccal cavity with three teeth, dorsal tooth double; 10 cephalic setae; double lateral cervical setae (usually); cuticle faintly striated; L = 2–8; p. 166.
Parironus, similar to *Thalassironus* except: caudal glands absent; L = 1.5–2.

F, *Trissonchulus***, buccal cavity with three teeth, dorsal tooth double; cephalic sensilla papilliform; tail short, blunt; L = 2–6; p. 168.
Pheronus, similar to *Trissonchulus* except: caudal glands absent; L = 1.5–2.

G, *Dolicholaimus***, three teeth, dorsal tooth single; cephalic sensilla papilliform; tail clavate; L = 2.5–4.5; p. 164.

H, *Syringolaimus*, three teeth; cephalic sensilla papilliform; posterior oesophageal bulb; tail long, narrow; L = 0.5–1.5.

I, ***Anoplostoma*****, buccal cavity large; teeth absent; copulatory bursa around cloaca; post-cloacal spine; L = 1–3; p. 150.

J, *Rhabdodemania***, buccal cavity with three teeth (usually); 4 + 6 cephalic setae; L = 3–8; p. 282.

K, *Pandolaimus*, buccal cavity large, divided into two sections; teeth absent; oesophagus tissue surrounds posterior half of buccal cavity; gubernaculum with apophysis; L = 1–2.

L, *Halanonchus*, anterior cephalic setae jointed; large oval structures surround large buccal cavity; amphids situated level with posterior of buccal cavity; 5–16 supplements in cervical region of male (usually); 7–17 precloacal supplements; tail elongated; L = 2–3.

M, ***Rhabditis*****, cephalic sensilla papilliform; buccal cavity tubular; oesophagus with bulb in middle; copulatory bursa around cloaca; L = 1–3; p. 294.

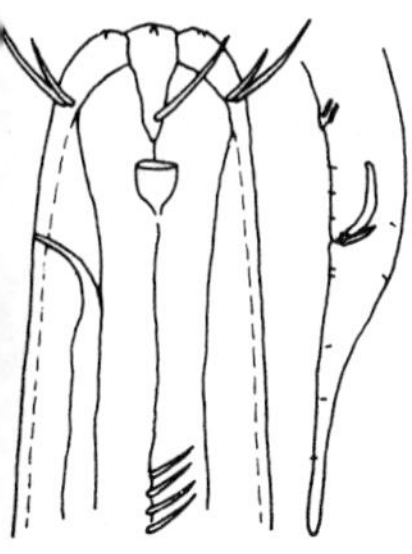

A. *Anticoma* 　B. *Paranticoma* 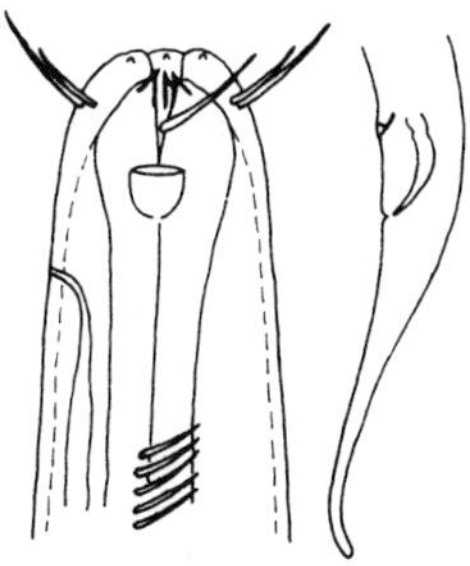　C. *Odontanticoma* 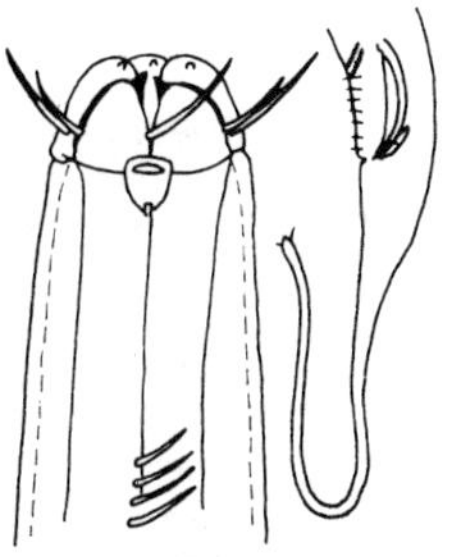　D. *Cephalanticoma*

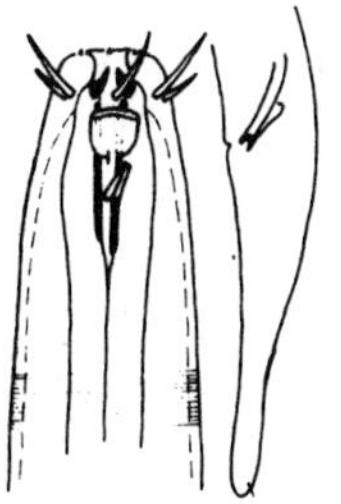

E. *Thalassironus* 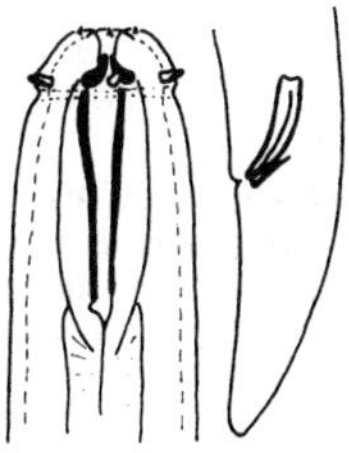　F. *Trissonchulus* 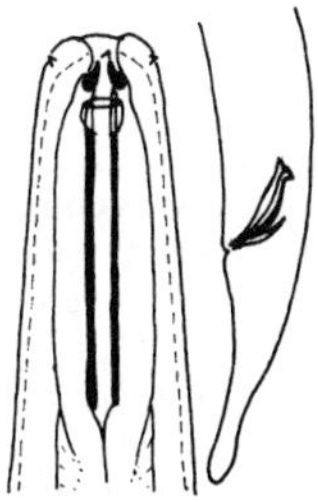　G. *Dolicholaimus* 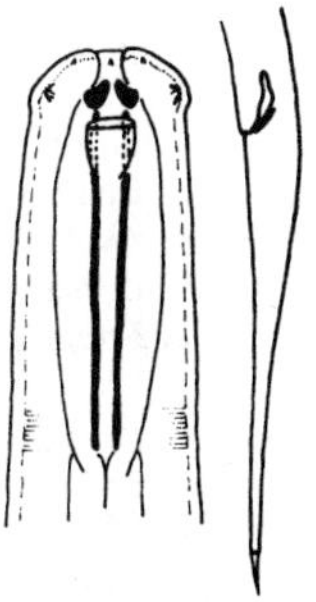　H. *Syringolaimus*

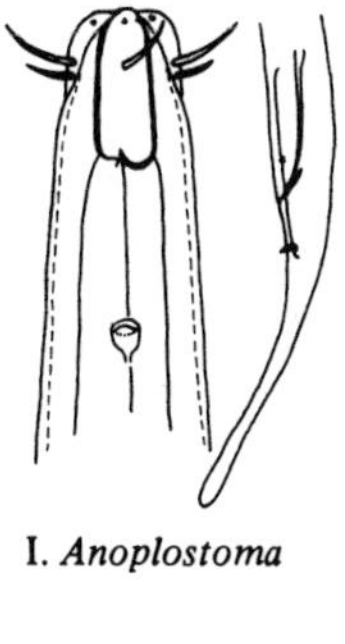

I. *Anoplostoma* 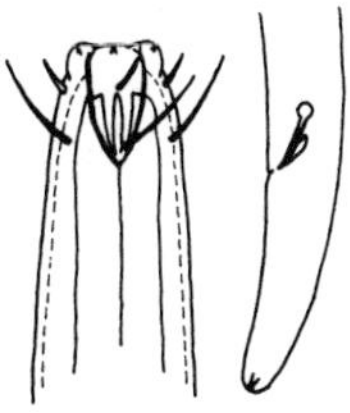　J. *Rhabdodemania* 　K. *Pandolaimus* 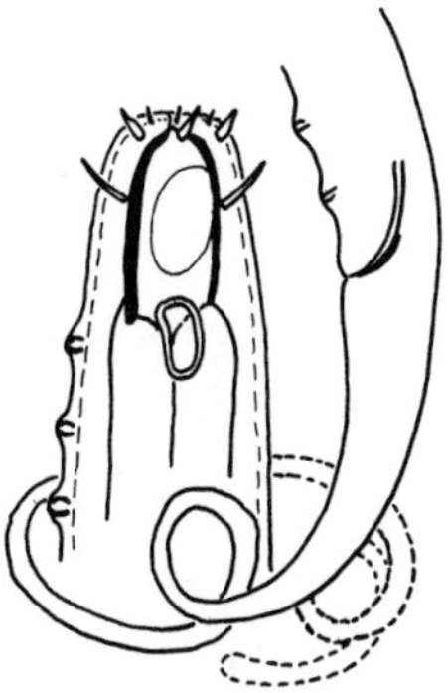　L. *Halanonchus*

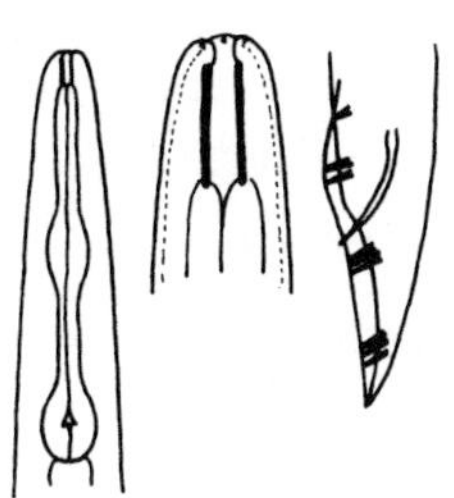

M. *Rhabditis*

Fig. 19.

A, ***Thoracostomopsis*****, buccal cavity with three elongate onchia forming a spear-like structure; three mandibles around anterior part of buccal cavity; cephalic capsule present; precloacal supplement present; L = 5–10; p. 146.

B, *Trileptium***, buccal cavity with three small mandibles and three onchia; precloacal supplement (usually); L = 2–8; p. 148.

C, *Gairleanema***, cephalic capsule absent; large labial setae with bulbous swellings; dorsal tooth; precloacal supplements; L = 2; p. 276.

D, *Chaetonema***, buccal cavity vase-shaped; teeth absent; male amphid an elongated groove; ventral precloacal spine; two pairs postcloacal spines; L = 1–2; p. 152.

E, ***Enoplolaimus*****, lips high; labial sensilla setiform; cephalic setae at posterior edge of cephalic capsule; three hollow mandibles; three teeth not extending anterior to mandibular bar; precloacal supplement (usually); L = 1–5; p. 120.

F, *Mesacanthion***, cephalic setae at middle or anterior of cephalic capsule; L = 2–5; p. 136.

G, *Paramesacanthion***, cephalic setae at anterior of cephalic capsule, level with anterior end of mandibles; cephalic setae much shorter in female than male; double articulated spicules; L = 2–3; p. 142.

H, *Oxyonchus*, two large subventral teeth extending to anterior bar of mandibles; dorsal tooth reduced; gubernaculum with apophysis; L = 2–4.

I, ***Enoplus*****, lips low; labial sensilla papilliform; mandibles solid; teeth absent; eyespots present or absent; precloacal supplement; L = 4–10; p. 102.

J, *Enoploides***, lips high, striated; solid mandibles with claw-like anterior; spicules usually long; L = 2–7; p. 112.

K, *Mesacanthoides***, lips unstriated; mandibles solid; L = 2–6; p. 136.

L, *Epacanthion***, mandibles broad, appearing as two longitudinal rods with a clear sheet of cuticle between; spicules long or short; L = 2–4; p. 130.

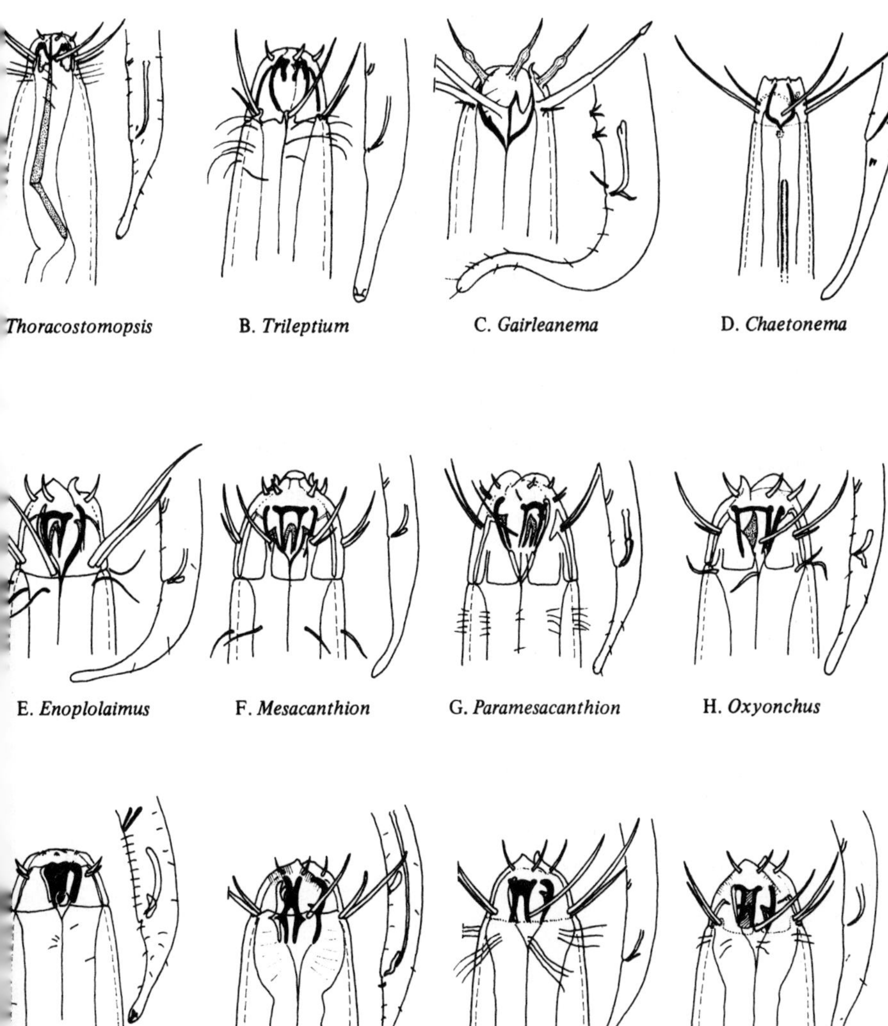
Thoracostomopsis
B. Trileptium
C. Gairleanema
D. Chaetonema
E. Enoplolaimus
F. Mesacanthion
G. Paramesacanthion
H. Oxyonchus
I. Enoplus
J. Enoploides
K. Mesacanthoides
L. Epacanthion

Fig. 20.

A, ***Viscosia*****, buccal cavity large with three teeth; right subventral tooth large; gubernaculum absent; female didelphic; L = 1–3; p. 236.

B, *Oncholaimellus***, right subventral tooth large and solid; buccal cavity divided transversely by cuticularised band; spicules long, equal or unequal; copulatory bursa present or absent; L = 1–2; p. 216.

C, *Adoncholaimus***, similar to *Viscosia* except: spicules long; gubernaculum present or absent; demanian system present in female; L = 2–7; p. 204.

D, ***Oncholaimus*****, left subventral tooth large; spicules short; gubernaculum absent; female monodelphic, demanian system present; tail short; L = 3–6; p. 220.

E, *Metoncholaimus***, similar to *Oncholaimus* except: spicules long (usually); gubernaculum present (usually); L = 3–8; p. 212.

F, *Prooncholaimus*, large bubble-like cells in body cavity between intestine and longitudinal chords (see tail picture); gubernaculum present; L = 2–4.

G, ***Pontonema*****, both subventral teeth large; gubernaculum present; single precloacal supplement; female didelphic, demanian system absent; L = 6–20; p. 232.

H, *Filoncholaimus*, similar to *Pontonema* except: filiform tail; L = 6–11.

I, *Metaparoncholaimus***, similar to *Oncholaimus* except: both subventral teeth large; L = 2–5; p. 210.

J, *Meyersia*, similar to *Viscosia* except: subventral teeth equal, dorsal tooth smaller; gubernaculum present; L = 3–6.

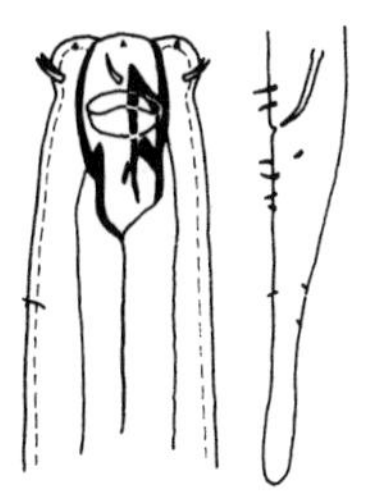

A. *Viscosia*

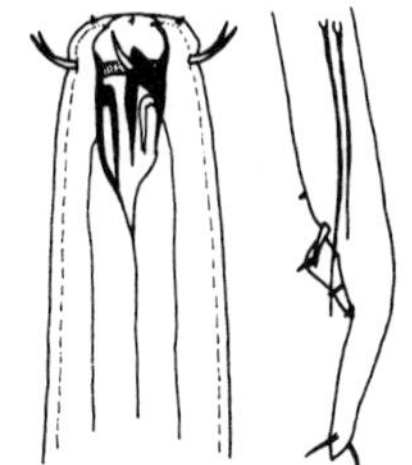

B. *Oncholaimellus*

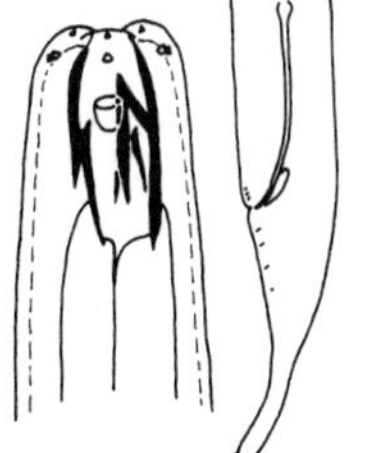

C. *Adoncholaimus*

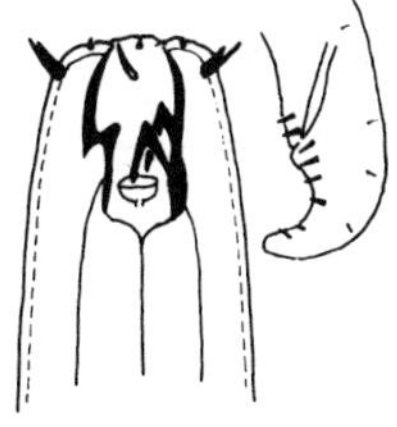

D. *Oncholaimus*

E. *Metoncholaimus*

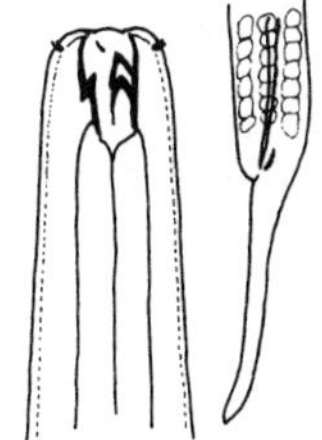

F. *Prooncholaimus*

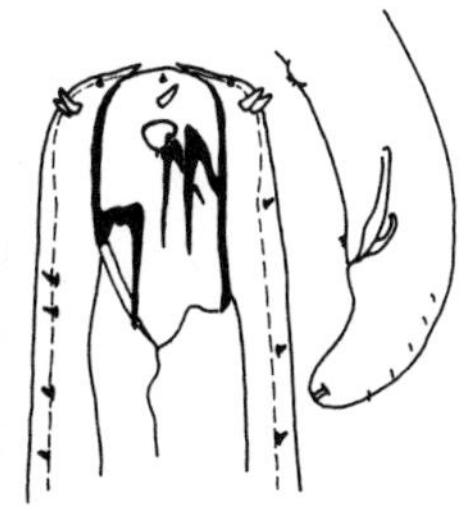

G. *Pontonema*

H. *Filoncholaimus*

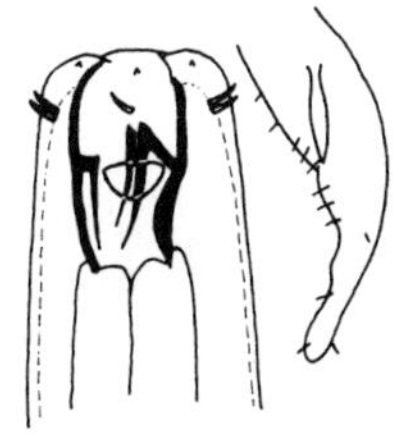

I. *Metaparoncholaimus*

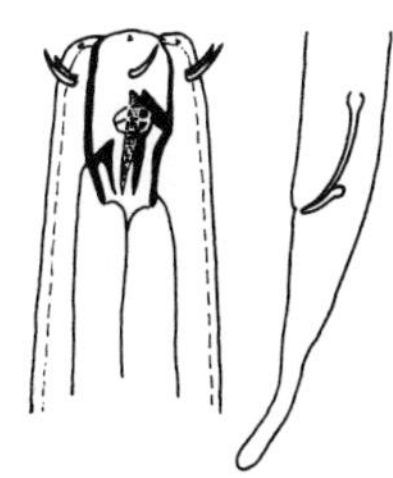

J. *Meyersia*

Fig. 21.

A, ***Ditlevsenella***, buccal cavity large with three teeth; right subventral tooth large; gubernaculum with apophysis; tail conical; L = 3–8.
B, ***Eurystomina*****, three teeth, right subventral large; 1–5 rows buccal denticles; two winged precloacal supplements; gubernaculum with apophysis; caudal glands present; L = 3–7; p. 252.
C, *Pareurystomina***, tail conical with sharp tip; caudal glands absent; L = 4–8; p. 258.
D, *Bathyeurystomina*, buccal cavity with band of denticles; large right subventral tooth only; two tubular precloacal supplements; tail conico-cylindrical; caudal glands absent; L = 7–11.
E, ***Belbolla*****, 7–10 oesophageal bulbs; winged precloacal supplements present or absent; L = 3; p. 248.
F, *Polygastrophora*, buccal cavity divided into several chambers; 5–8 oesophageal bulbs; precloacal supplements absent; L = 2–3.
G, ***Calyptronema*****, buccal cavity sexually dimorphic, female divided into two chambers, male absent; female oesophagus has dilated cuticularised lumen; tail long or short; precloacal papillae present; L = 2–5; p. 250.
H, *Symplocostoma***, female buccal cavity divided into several chambers, male buccal cavity absent; precloacal papillae present or absent; L = 3–6; p. 262.

A. *Ditlevsenella*

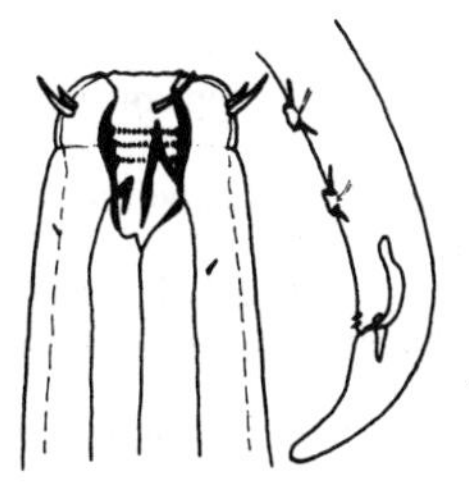
B. *Eurystomina*

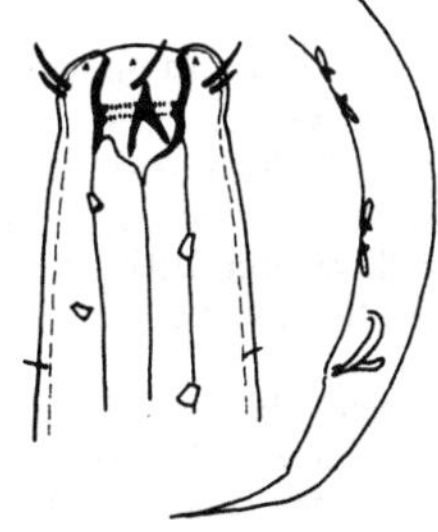
C. *Pareurystomina*

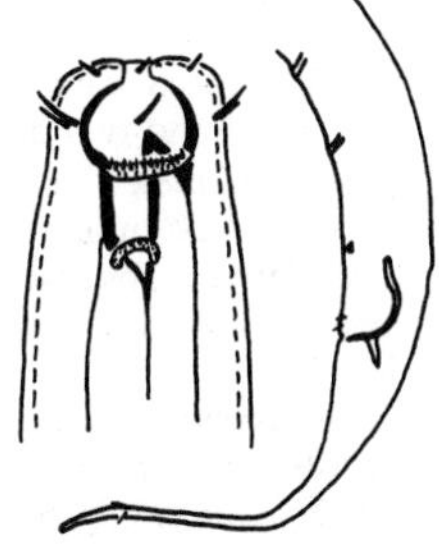
D. *Bathyeurystomina*

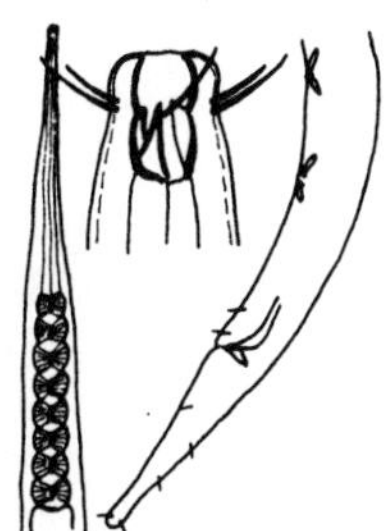
E. *Belbolla*

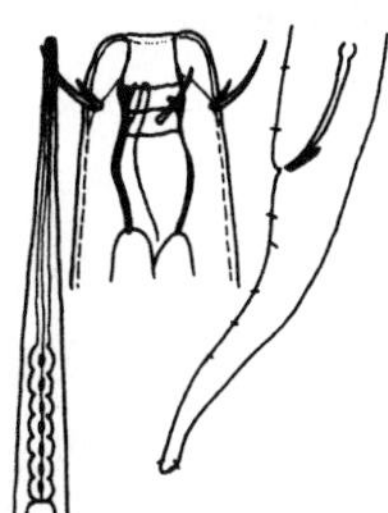
F. *Polygastrophora*

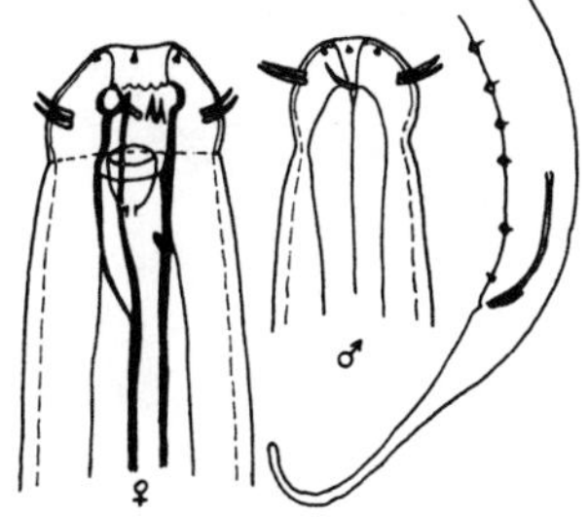

G. *Calyptonema*

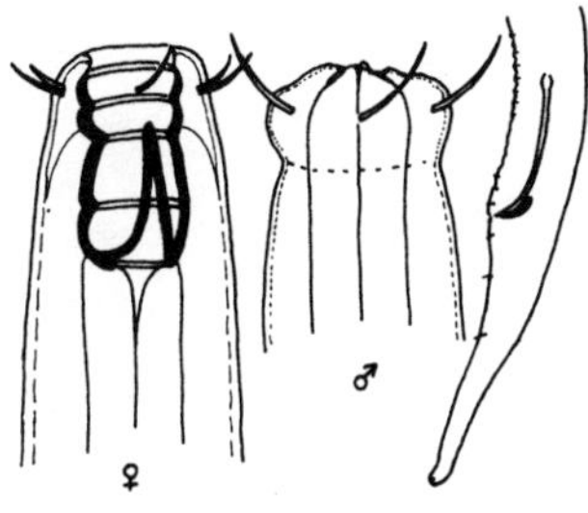

H. *Symplocostoma*

Fig. 22.

A, ***Leptolaimus****, buccal cavity long, cylindrical; four cephalic setae; amphids circular; 2–16 tubular precloacal supplements; L = 0.5–1.
*Halaphanolaimus**, similar to *Leptolaimus* except: preanal supplements present in female also; L = 1–2.

B, *Stephanolaimus**, buccal cavity small, conical; amphids small; six anterior cephalic sensilla may also be elongated; 9–45 tubular precloacal supplements (may be present in female also); L = 1–3.

C, *Antomicron**, amphids an elongated loop; 3–10 tubular precloacal supplements; L = 0.5–1.5.

D, *Leptolaimoides*, buccal cavity long, cylindrical; four short cephalic setae; amphids elongated; tubular precloacal supplements present or absent; tail with filiform posterior section; L = 0.5–1.

E, ***Dagda****, buccal cavity with three teeth; 10–14 tubular precloacal supplements; one precloacal and two postcloacal papillae; L = 2–4.

F, *Diodontolaimus**, buccal cavity cylindrical; two subventral teeth; one precloacal papilla; 9–20 tubular precloacal supplements; L = 1–2.

G, *Cricolaimus**, buccal cavity small, surrounded by cuticularised ring with dorsal and smaller ventral denticles; 13–17 tubular precloacal supplements, in two groups; L = 2–4.

H, *Lauratonema*, buccal cavity conical; 10 cephalic setae; gubernaculum absent or indistinct; precloacal supplements absent; L = 1–2.

I, ***Camacolaimus****, buccal cavity with long stylet-like dorsal tooth, not always easy to detect; amphid spiral; precloacal supplements absent; tail tip pointed; L = 1–3.

J, *Onchium**, buccal cavity with stylet; ocellus present; L = 0.5–2.

K, *Procamacolaimus*, buccal cavity with stylet; 5–17 precloacal supplements; L = 1–2.

L, *Deontolaimus*, 17–40 oesophageal supplements; L = 0.5–1.

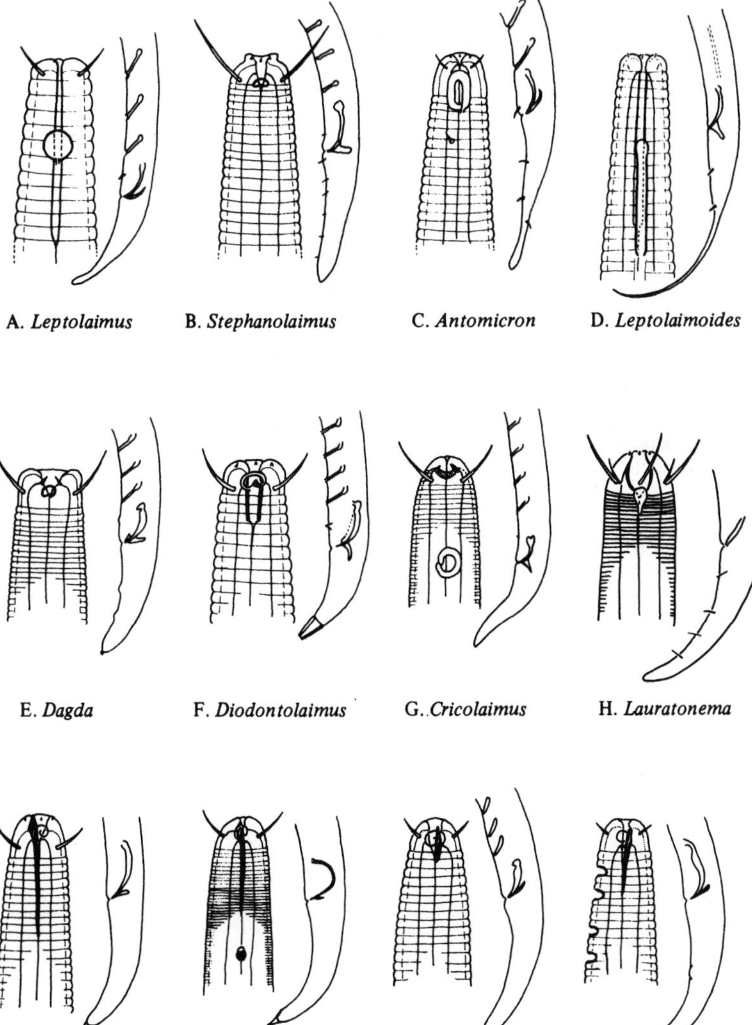

A. *Leptolaimus* B. *Stephanolaimus* C. *Antomicron* D. *Leptolaimoides*

E. *Dagda* F. *Diodontolaimus* G. *Cricolaimus* H. *Lauratonema*

I. *Camacolaimus* J. *Onchium* K. *Procamacolaimus* L. *Deontolaimus*

Fig. 23.

A, ***Axonolaimus****, buccal cavity double-conical; amphid an elongated loop; female didelphic; L = 1–2.

B, *Ascolaimus**, amphid a rounded loop; tail slightly swollen at the posterior end; L = 2–6.

C, ***Odontophora****, buccal cavity with a ring of six teeth in the anterior part (odontia); L = 1–4.

D, *Synodontium*, similar to *Axonolaimus* except: lateral subcephalic seta present; female monodelphic; L = 2–5.

Apodontium, similar to *Synodontium* except: amphids indistinct; female didelphic; L = 2–3.

Odontophoroides, similar to *Odontophora* except: lateral subcephalic seta present; female monodelphic; L = 1–1.5.

E, *Parodontophora*, walls of buccal cavity parallel; L = 1–2.

F, ***Araeolaimus****, buccal cavity narrow, conical; amphid a rounded loop; oesophagus with a bulbous swelling midway; ocellus present, but may be lost on fixation; L = 1–3.

G, *Southerniella*, buccal cavity cylindrical; amphid circular; set-off anterior part of oesophagus (corpus); L = 0.5–2.

H, *Diplopeltis**, buccal cavity absent; amphids on plaques; ocellus present; L = 2–4.

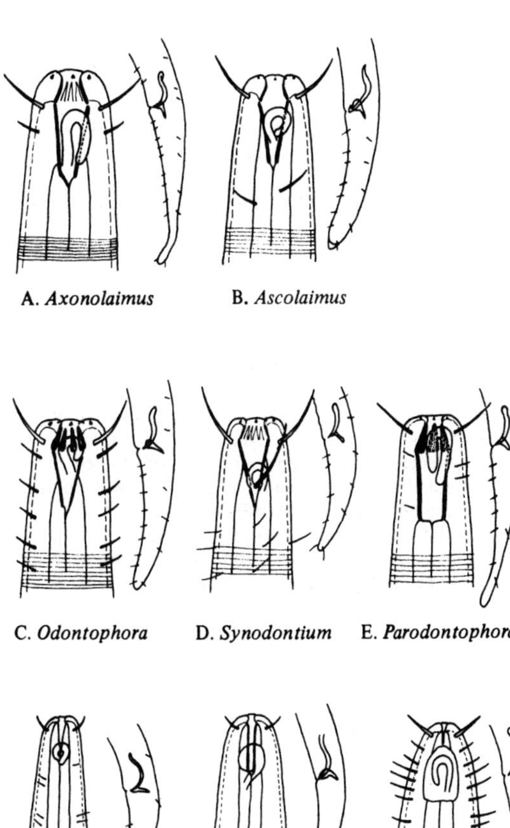

A. *Axonolaimus* B. *Ascolaimus*

C. *Odontophora* D. *Synodontium* E. *Parodontophora*

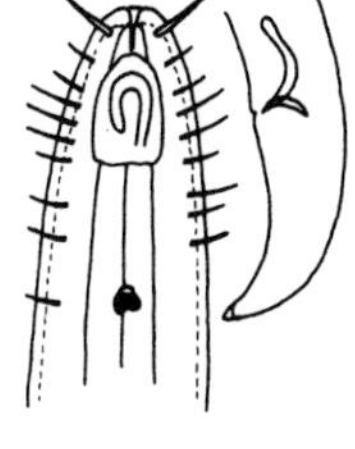

F. *Araeolaimus* G. *Southerniella* H. *Diplopeltis*

Fig. 24.

A, ***Campylaimus****, buccal cavity subterminal; ventral limb of amphid extends to tail; L = 0.5–1.5.

B, *Pseudolella*, buccal cavity large; ventral limb of amphid elongated; L = 0.5–1.5.

C, *Tarvaia**, buccal cavity absent; amphid large, spiral and situated on plaque; L = 0.7–2.5.

D, ***Diplopeltula****, buccal cavity small or absent, terminal or subterminal; amphid on plaque; L = 1–2.

E, *Coninckia*, buccal cavity absent; 6 + 4 cephalic setae; amphid on plaque (usually); L = 1–2.

F, *Chitwoodia**, buccal cavity absent; 6 + 4 cephalic setae; L = 1–2.

G, *Tubolaimoides*, buccal cavity absent; 6 + 4 cephalic setae; amphid circular; L = 1–2.

H, ***Cyartonema****, buccal cavity absent; amphid circular; oesophagus with narrow isthmus; L = 0.5–1.5.

I, *Aegialoalaimus*, buccal cavity long, narrow; amphid circular; L = 0.5–1.5.

J, *Siphonolaimus**, buccal cavity with axial spear; L = 4–10.

K, *Diplopeltoides*, buccal cavity absent; amphid loop-shaped; L = 0.5.

L, ***Haliplectus****, cephalic setae absent; amphid circular; mid-oesophageal bulb; precloacal supplements (usually); L = 0.5–2.

M, *Setoplectus*, similar *Haliplectus* except: four cephalic setae present; L = 1.

N, *Rhabditis***, cephalic sensilla papilliform; amphid absent; buccal cavity tubular; copulatory bursa around cloaca; L = 1–3; p. 294.

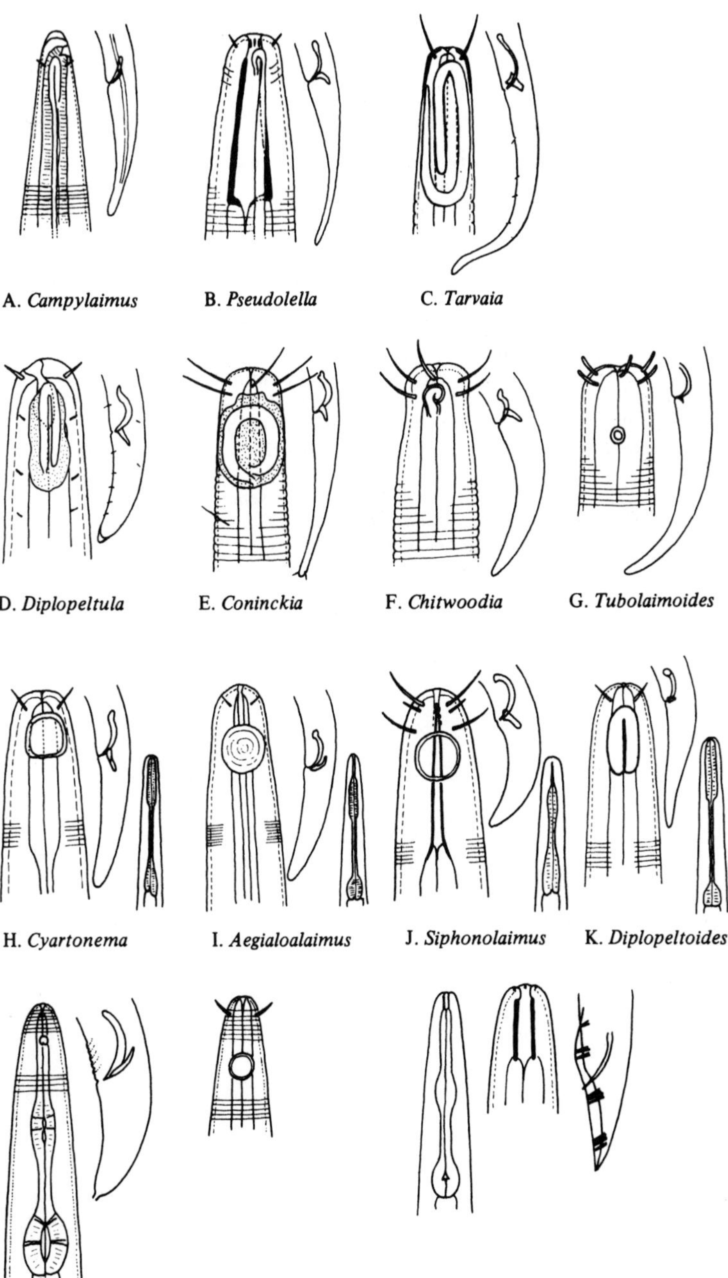
A. Campylaimus
B. Pseudolella
C. Tarvaia
D. Diplopeltula
E. Coninckia
F. Chitwoodia
G. Tubolaimoides
H. Cyartonema
I. Aegialoalaimus
J. Siphonolaimus
K. Diplopeltoides
L. Haliplectus
M. Setoplectus
N. Rhabditis

Fig. 25.

A, ***Metalinhomoeus****, buccal cavity small; four cephalic setae; four subcephalic setae; amphid circular; posterior oesophageal bulb present; L = 2–4.

B, *Terschellingia**, buccal cavity absent or only small; four cephalic setae; subcephalic setae present; L = 1–2.

C, ***Desmolaimus****, buccal cavity with transverse cuticular rings; two median subcephalic setae; faint cuticle striations; L = 1–3.

D, *Megadesmolaimus*, buccal cavity without cuticular rings; labial sensilla setiform; L = 2–5.

E, *Anticyathus*, amphid spiral; cephalic sensilla papilliform; L = 2–7.

F, ***Linhomoeus****, buccal cavity posteriorly heavily cuticularised with teeth and/or plates; 10 cephalic setae; tail cylindrical; L = 2–10.

G, *Paralinhomoeus**, buccal cavity unarmed and weakly cuticularised; 10 cephalic setae; tail tapered (usually); L = 1–4.

H, *Didelta*, amphid oval, on plaque; tail conico-cylindrical; L = 2–4.

I, *Disconema**, buccal cavity reduced; amphid oval; L = 1–4.

J, ***Eleutherolaimus****, buccal cavity cylindrical; 4 + 4 cephalic setae; L = 1.5–3.5.

K, *Eumorpholaimus**, buccal cavity deep, cylindrical; 4 + (2 + 4) cephalic setae; L = 2–4.

L, *Gnomoxyala*, buccal cavity cylindrical; amphid absent; four cephalic setae; cuticle only faintly striated; female monodelphic; L = 0.4–0.5.

M, *Retrotheristus*, similar to *Gnomoxyala* except: buccal cavity conical; L = 0.6–0.7.

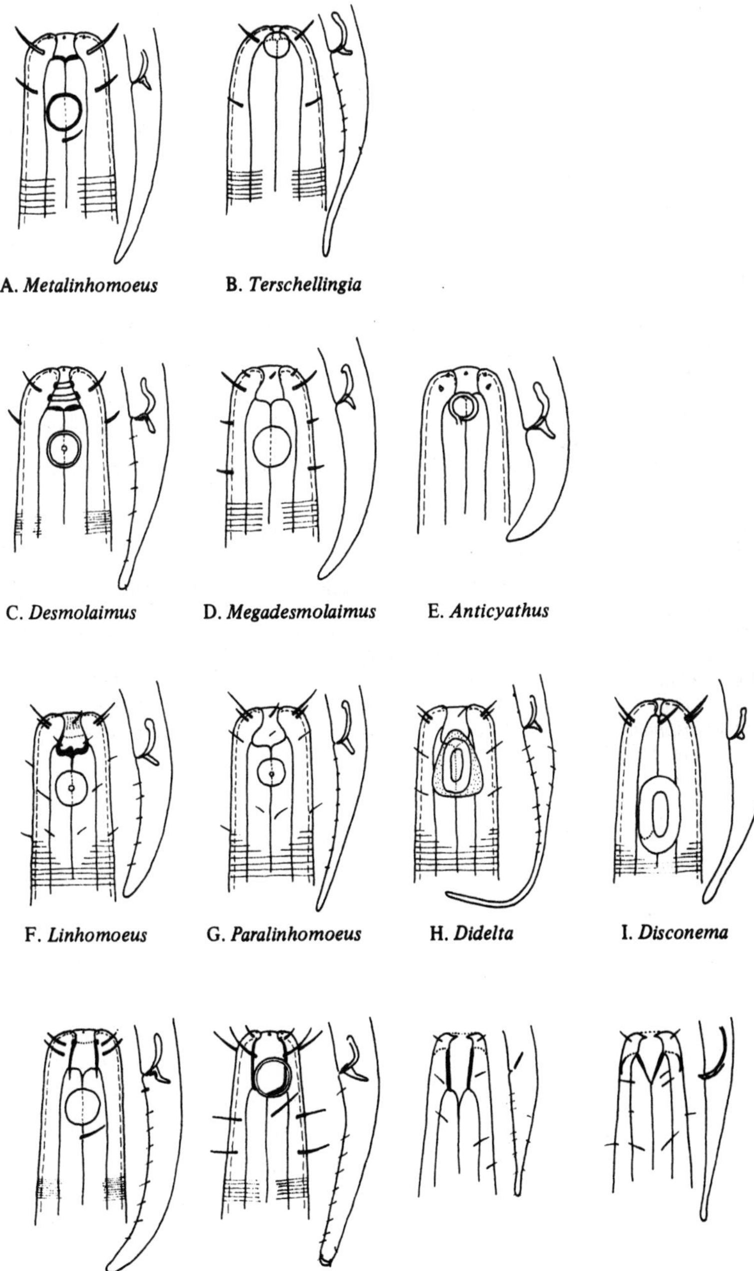

A. *Metalinhomoeus* B. *Terschellingia*

C. *Desmolaimus* D. *Megadesmolaimus* E. *Anticyathus*

F. *Linhomoeus* G. *Paralinhomoeus* H. *Didelta* I. *Disconema*

J. *Eleutherolaimus* K. *Eumorpholaimus* L. *Gnomoxyala* M. *Retrotheristus*

Fig. 26. *Note:* Females of all these genera are monodelphic.

A, ***Monhystera****, cuticle smooth; 6–10 cephalic setae in six groups; amphid circular; male monorchic; L = 0.5–3.
B, *Diplolaimella*, buccal cavity double; tail long, conico-cylindrical; precloacal and postcloacal papillae present or absent; L = 0.5–1.
C, *Diplolaimelloides*, buccal cavity double; amphid spiral; copulatory bursa present; L = 0.6–1.2.
D, *Gammarinema*, buccal cavity double, teeth in posterior part; commensal on Crustacea; L = 1–2.
E, ***Theristus****, buccal cavity conical; 10–14 cephalic setae in six groups; amphid circular; tail conical; L = 1–2.
F, *Metadesmolaimus**, buccal cavity extended anteriorly; cuticle brownish; L = 0.8–2.
G, *Ammotheristus*, 10 or more cephalic setae; amphid large and bladder-like; long somatic setae; tail conical; terminal setae absent; L = 0.5–1.
H, *Amphimonhystera**, labial sensilla setiform; amphid large with internal cuticularised opening; buccal cavity has a posterior section with three pharyngeal tubes; cuticle light brown; L = 1.5–2.
I, ***Daptonema****, buccal cavity conical; 10–14 cephalic setae in six groups; amphid circular; tail conico-cylindrical; L = 0.5–2.
J, *Paramonohystera**, similar to *Daptonema* except: spicules elongated; L = 1–2.
K, *Promonhystera*, similar to *Paramonohystera* except: labial sensilla setiform; tail conical or conico-cylindrical; L = 0.6–1.5.
L, *Stylotheristus*, labial sensilla conical; spicules short; oesophageal muscle around buccal cavity well developed; L = 2–2.5.
M, ***Steineria***, 10 cephalic setae, six long and four short; eight groups of long subcephalic setae at same level as cephalic setae; L = 1–2.
N, *Pseudosteineria*, eight groups of long cervical setae level with amphids; L = 1–2.
O, *Trichotheristus**, long somatic setae, especially in oesophagus region; L = 1–2.
P, *Subsphaerolaimus*, buccal cavity strongly cuticularised; eight groups of subcephalic setae; L = 0.8–1.

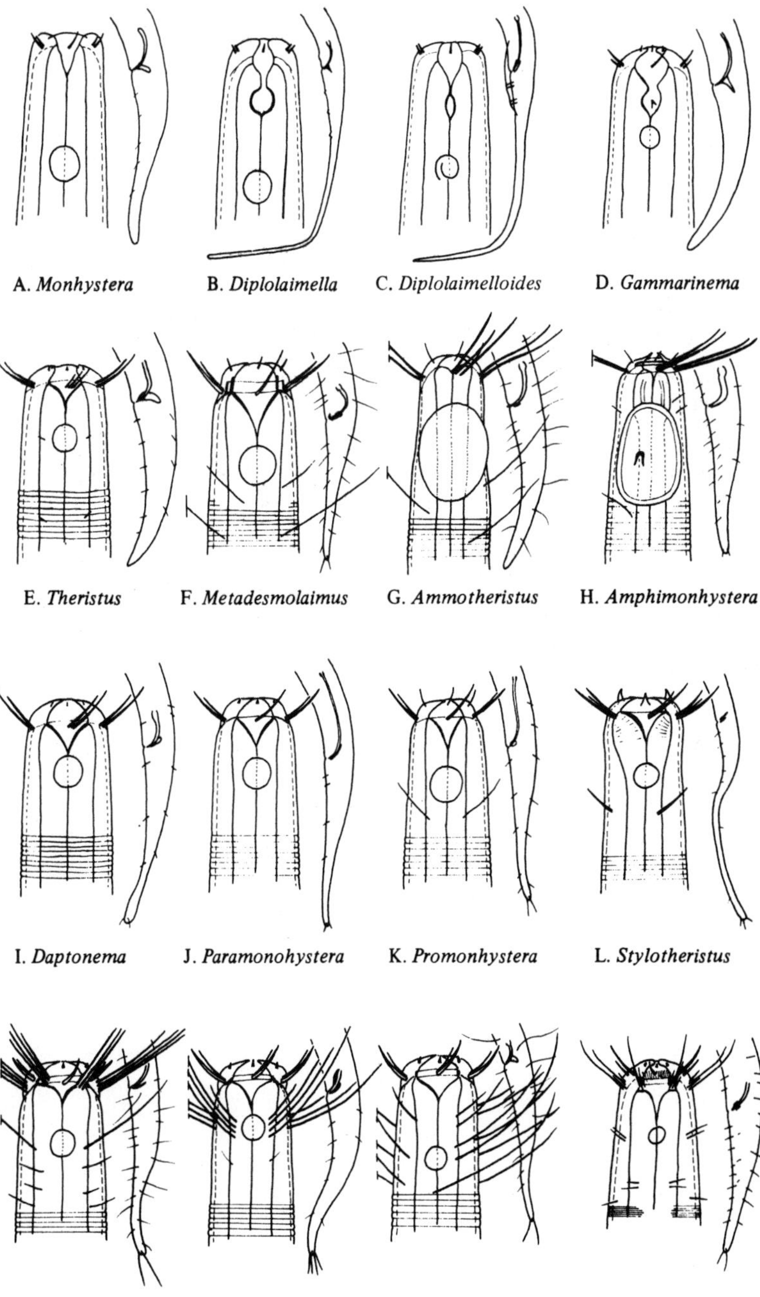
A. *Monhystera*
B. *Diplolaimella*
C. *Diplolaimelloides*
D. *Gammarinema*
E. *Theristus*
F. *Metadesmolaimus*
G. *Ammotheristus*
H. *Amphimonhystera*
I. *Daptonema*
J. *Paramonohystera*
K. *Promonhystera*
L. *Stylotheristus*
M. *Steineria*
N. *Pseudosteineria*
O. *Trichotheristus*
P. *Subsphaerolaimus*

Fig. 27. *Note:* Females of all these genera are monodelphic.

A, ***Linhystera***, buccal cavity minute; 10 cephalic setae; tail conico-cylindrical; L = 0.7–0.8.
B, ***Elzalia***, buccal cavity strongly cuticularised, cylindrical; spicules elongated; L = 0.5–1.
C, *Amphimonhystrella*, buccal cavity deep, conical; spicules short; L = 0.5.
D, *Gnomoxyala*, buccal cavity cylindrical; amphid absent; four cephalic setae; cuticle only faintly striated; L = 0.4–0.5.
E, *Retrotheristus*, similar to *Gnomoxyala* except buccal cavity conical; L = 0.6–0.7.
F, ***Cobbia****, buccal cavity with three teeth; labial sensilla setiform; tail filiform; L = 1–2.
G, *Valvaelaimus*, buccal cavity with three tooth-like processes; amphid with cuticularised internal opening; spicules elongated; cuticle light brown; L = 2–3.5.
H, *Rhynchonema*, buccal cavity elongated; anterior third of oesophagus region attenuated; amphid posterior to attenuated section; L = 0.4–1.
I, *Scaptrella*, buccal cavity with six eversible jointed odontia; tail long or short; L = 1–2.
J, ***Sphaerolaimus****, buccal cavity large, barrel-shaped, heavily cuticularised; inner oesophagus wall cuticularised; 10 cephalic setae (six short, four long); eight groups of subcephalic setae; two testes; L = 1–4.
K, *Parasphaerolaimus**, buccal cavity conical with six anteriorly situated plates; inner oesophagus wall not especially cuticularised; single testis; L = 1.5–3.
L, *Doliolaimus*, buccal cavity cylindrical, wide; eight subcephalic setae; L = 1.
M, *Subsphaerolaimus*, buccal cavity short; eight groups of subcephalic setae; L = 0.8–1.

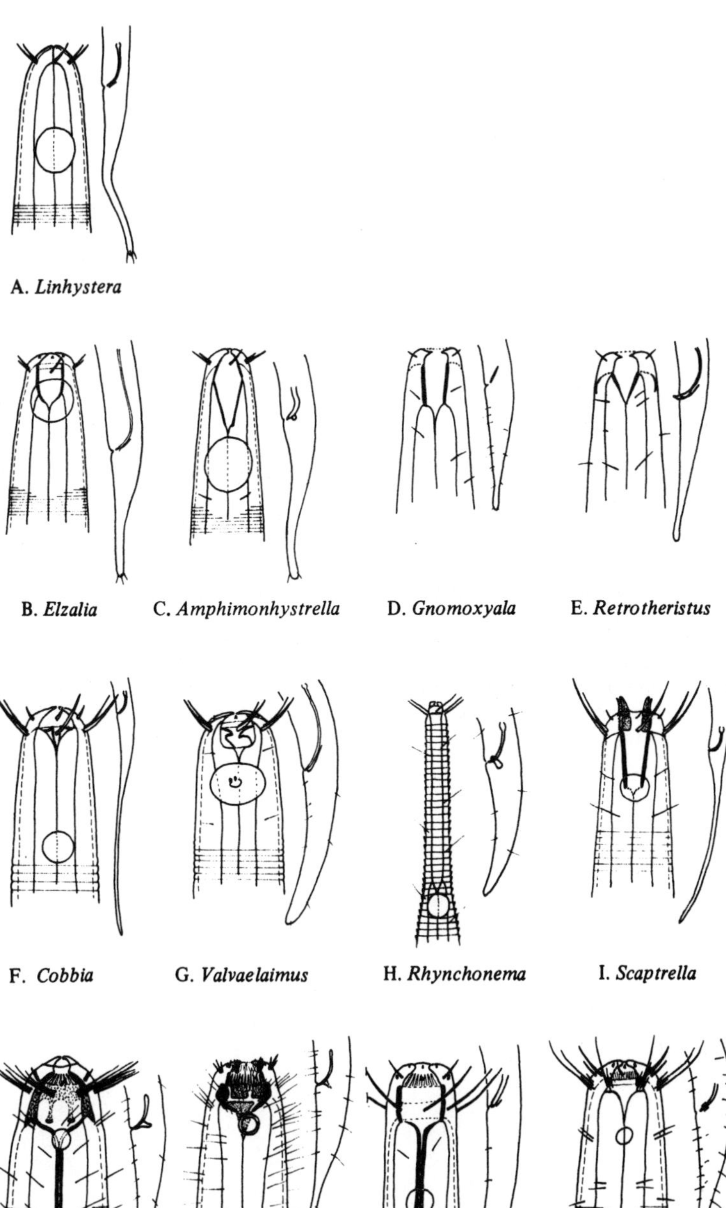

A. *Linhystera*

B. *Elzalia* C. *Amphimonhystrella* D. *Gnomoxyala* E. *Retrotheristus*

F. *Cobbia* G. *Valvaelaimus* H. *Rhynchonema* I. *Scaptrella*

J. *Sphaerolaimus* K. *Parasphaerolaimus* L. *Doliolaimus* M. *Subsphaerolaimus*

Fig. 28.

A, ***Xyala****, cuticle with longitudinal ridges; labial sensilla setiform; single ovary; L = 1–2.
B, *Gonionchus**, lips hyaline and usually extended anteriorly; longitudinal ridges present; tail rather long; L = 1–2.
C, *Omicronema*, cuticle with transverse rows of longitudinal striations; buccal cavity large, cylindrical; amphid contour indistinct; L = 1–2.
D, *Echinotheristus*, cuticle with transverse rows of small spines; amphid bladder-like with indistinct contour; precloacal supplements present; L = 0.7–1.5.
E, ***Monoposthia****, cuticle with longitudinal ridges (alae) of V-shaped structures; four cephalic setae; elongated oesophageal bulb; spicules absent; single gubernaculum; single testis; single ovary; L = 1–3.
F, *Nudora**, similar to *Monoposthia* except: spicules present; L = 1–2.
Monoposthoides, similar to *Monoposthia* except: alae absent in the anterior part of the oesophagus; two testes; L = 1–2.
G, *Rhinema*, alae point in the opposite direction to those of *Monoposthia*, spicules present; two ovaries; L = 1.
H, *Xenolaimus*, head set-off; 10 cephalic setae; single ovary; L = 1–2.
I, ***Ceramonema****, 70–250 wide body annules and eight longitudinal ridges; buccal cavity absent; 6 + 4 cephalic setae; amphid loop-shaped; two ovaries; L = 0.5–2.
J, *Dasynemoides**, over 500 body annules; cephalic capsule elongated; L = 1–2.
K, *Metadasynemoides*, cephalic capsule short; 10 separate cephalic setae in one circle; L = 1–2.
L, *Metadasynemella*, about 100 body annules; cuticle vacuolated below the longitudinal ridges; 10 cephalic setae; L = 0.5.
M, ***Pselionema****, 70–350 body annules; buccal cavity absent; four cephalic setae; L = 0.4–1.5.
N, *Pterygonema**, longitudinal ridges conspicuously extended; about 320 body annules; cuticular spine between limbs of amphid; gubernaculum with caudal apophysis; L = 0.8–1.

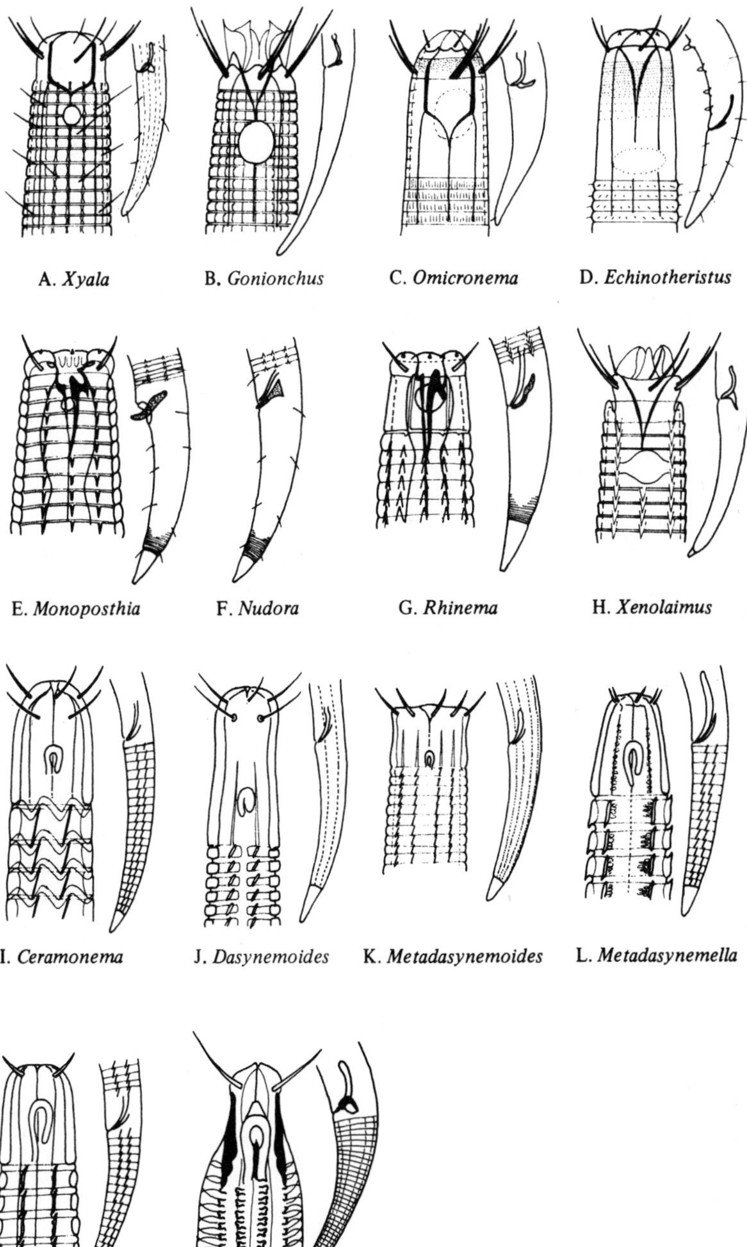

A. *Xyala* B. *Gonionchus* C. *Omicronema* D. *Echinotheristus*

E. *Monoposthia* F. *Nudora* G. *Rhinema* H. *Xenolaimus*

I. *Ceramonema* J. *Dasynemoides* K. *Metadasynemoides* L. *Metadasynemella*

M. *Pselionema* N. *Pterygonema*

Fig. 29.

A, ***Gerlachius****, short stout body; cuticle faintly striated, may look smooth; four cephalic setae on peduncles; amphid round, sometimes indistinct; ovaries reflexed; L = 0.5.

B, *Meylia*, cuticle with rows of pegs or spines; amphid spiral; ovaries outstretched; L = 0.6–1.2.

C, ***Microlaimus****, head set-off; six short and four long cephalic setae; amphid surrounded by cuticular striations; oval oesophageal bulb; tail conical; L = 0.5–2.

D, *Calomicrolaimus**, cervical region elongated; coarse ventral striations anterior to cloaca; L = 0.5–2.

E, *Molgolaimus**, similar to *Microlaimus* except: four cephalic setae; spherical oesophageal bulb; tail conico-cylindrical; spicules may be elongated; L = 0.8–1.5.

F, *Aponema*, similar to *Molgolaimus* except: gubernaculum with apophysis; L = 0.6–0.7.

G, ***Paramicrolaimus***, 6 + 4 long cephalic setae; oesophageal bulb only slight; pre-cloacal supplements present or absent; L = 2–5.

H, *Bolbolaimus**, head not set-off; pharyngeal bulb present; L = 1–2.

I, *Ixonema*, cuticle smooth, covered with sediment; amphid with elongated corpus gelatum; L = 0.5.

J, ***Spirinia****, buccal cavity with minute (or absent) dorsal tooth; four cephalic setae; amphid surrounded by cuticle striations; L = 2–4.

K, *Chromaspirina**, buccal cavity with well-developed dorsal tooth; 6 + 4 cephalic setae; oesophageal bulb only slight; L = 1–4.

L, *Parallelocoilas*, buccal cavity large, cylindrical with dorsal tooth; four subcephalic setae level with four cephalic setae; oesophageal bulb absent; L = 0.6–0.7.

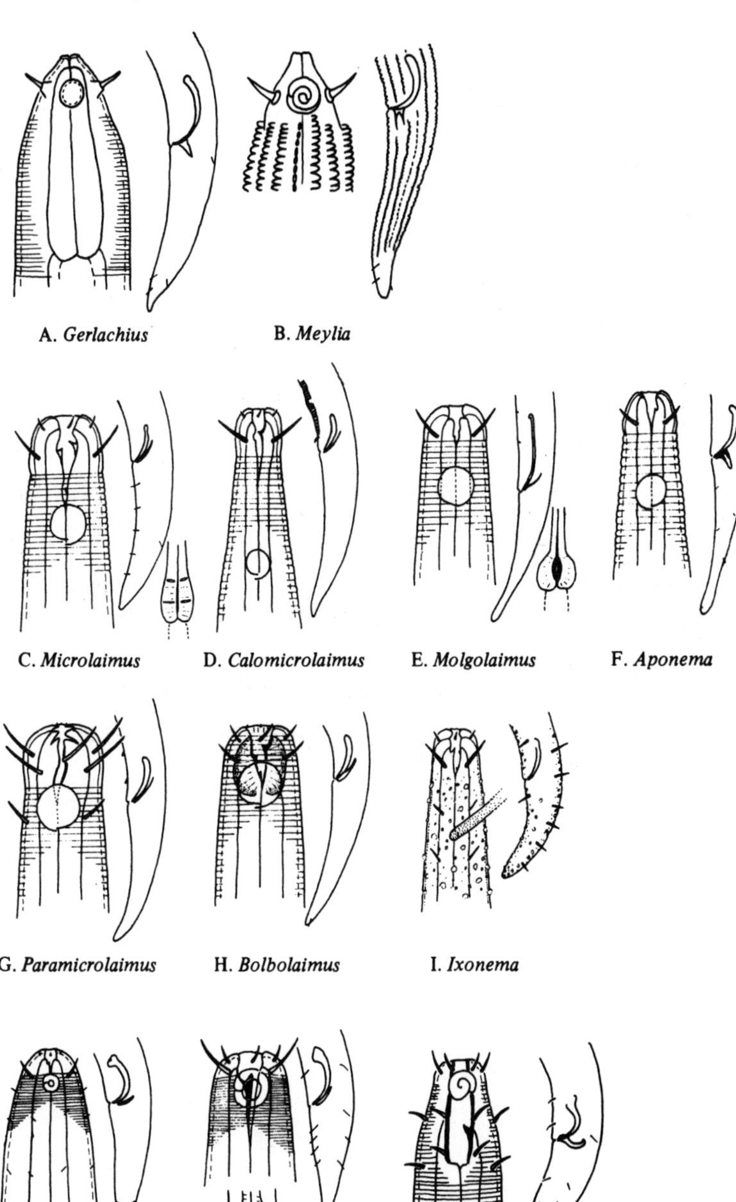
A. *Gerlachius*
B. *Meylia*
C. *Microlaimus*
D. *Calomicrolaimus*
E. *Molgolaimus*
F. *Aponema*
G. *Paramicrolaimus*
H. *Bolbolaimus*
I. *Ixonema*
J. *Spirinia*
K. *Chromaspirina*
L. *Parallelocoilas*

Fig. 30.

A, ***Metachromadora****, buccal cavity with dorsal tooth; oesophageal bulb well developed with thick cuticular lining; indistinct precloacal supplements (usually); L = 1–2.

B, *Metachromadora**, some species have cuticularised precloacal supplements extending almost to the oesophagus.

C, *Metadesmodora*, amphid on plaque; numerous somatic setae; L = 1–2.

D, ***Sigmophoranema****, buccal cavity with dorsal tooth; six short and four long cephalic setae; eight subcephalic setae; amphid anterior to cuticular striations; spicules elongated; S-shaped precloacal supplements; L = 1–2.

E, *Onyx**, buccal cavity with dorsal tooth; spicules short; L = 1–2.

F, *Polysigma*, similar to *Sigmophoranema* except: subcephalic setae absent; spicules short; L = 1–2.

G, *Pseudonchus**, buccal cavity in two parts, anterior with rows of odontia at the anterior margin; precloacal supplements present or absent; L = 1.5–3.

H, ***Desmodora****, head set-off with conspicuous cephalic capsule; four cephalic setae; amphid anterior to cuticular striations; tubular precloacal supplements present; L = 1–10.

Acanthopharynx, similar to *Desmodora* except: oesophageal bulb very long, almost half length of oesophagus; L = 1–3.

I, *Paradesmodora*, amphids half surrounded by cuticular striations; precloacal supplements present; L = 1–3.

J, ***Leptonemella****, buccal cavity absent; four cephalic setae; eight subcephalic setae; male amphids loop-shaped; cuticle covered with epizoites in life but may be lost on fixation; L = 2–4.

K, *Eubostrichus*, buccal cavity absent; head not set-off; two pairs of cervical subventral hollow spines; precloacal and postcloacal subventral hollow spines; L = 2–3.

L, *Catanema**, four cephalic setae; subcephalic setae present; muscular pharyngeal bulb; gubernaculum with apophysis; L = 2–5.

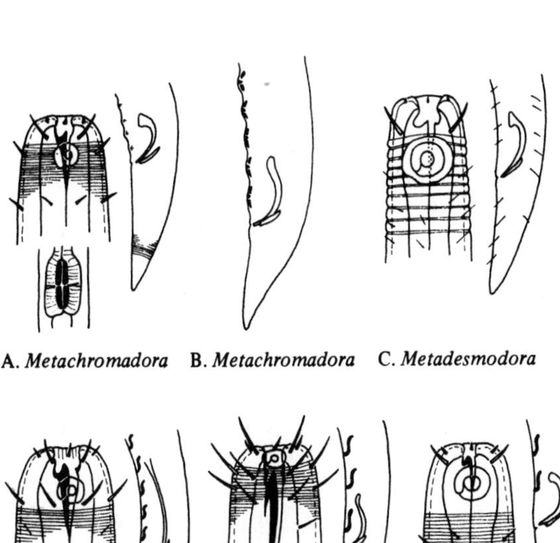

A. *Metachromadora* B. *Metachromadora* C. *Metadesmodora*

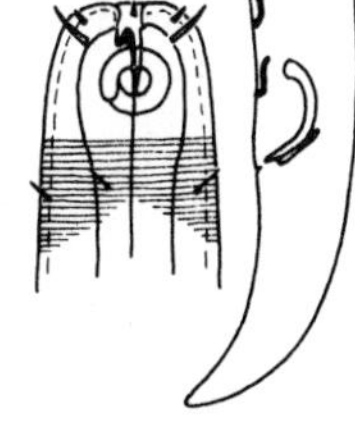

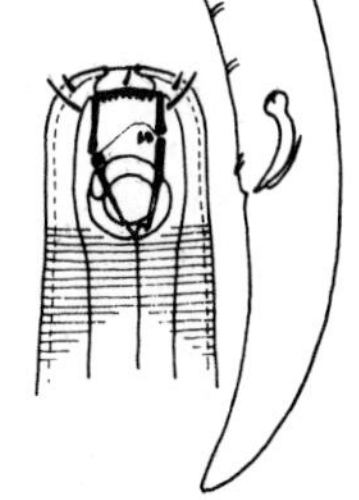

D. *Sigmophoranema* E. *Onyx* F. *Polysigma* G. *Pseudonchus*

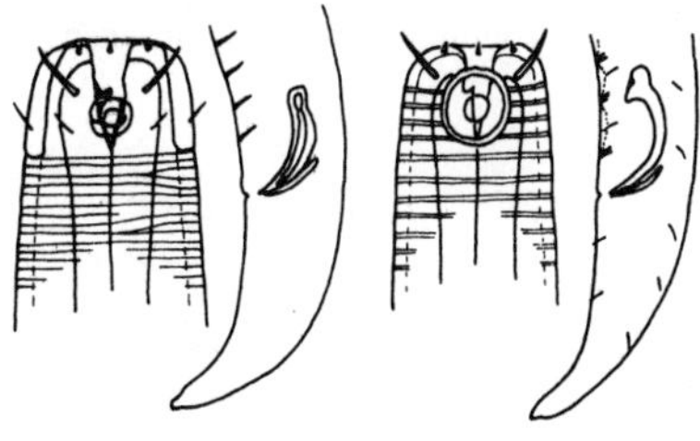

H. *Desmodora* I. *Paradesmodora*

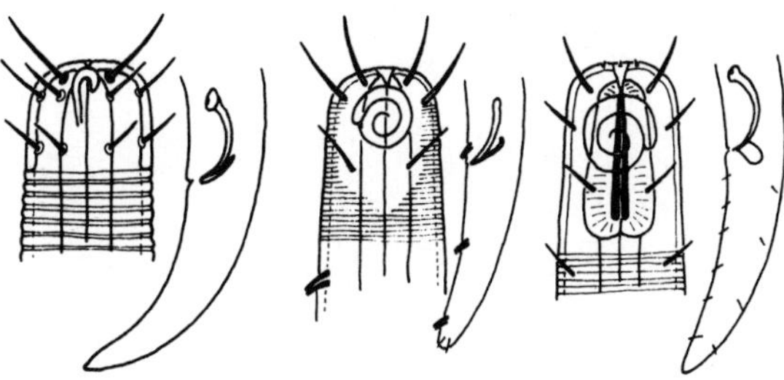

J. *Leptonemella* K. *Eubostrichus* L. *Catanema*

Fig. 31.

A, ***Sabatieria****, buccal cavity cup-shaped, teeth absent; 6 + 4 cephalic setae; amphid multi-spiral; cuticle punctated but may be difficult to detect; lateral differentiation present or absent but longitudinal rows are absent; gubernaculum with dorsal apophysis; precloacal supplements present (usually); L = 1–3.

B, *Pierrickia*, buccal cavity small; 6 + 4 short cephalic setae; cuticle with lateral differentiation of irregular dots but no longitudinal rows; precloacal supplements absent; L = 1.

C, *Laimella**, buccal cavity with three small teeth; cuticle transversely striated; L = 0.5–2.

D, *Cervonema*, buccal cavity small, no teeth; cervical region elongated; amphid situated at or more than 2 h.d. from anterior; L = 1–1.5.

E, ***Dorylaimopsis****, buccal cavity cylindrical, three teeth present; 6 + 4 cephalic setae; cuticle with lateral differentiation of longitudinal rows of dots; spicules elongated, often appearing as a double bow shape; gubernaculum apophysis directed caudally; precloacal supplements present; L = 2–2.5.

F, *Hopperia*, cuticle with lateral differentiation of irregular dots but no longitudinal rows; spicules short; L = 2–2.5.

G, *Paramesonchium*, buccal cavity conical, three teeth present; L = 2.5.

H, *Vasostoma*, buccal cavity cylindrical with three teeth; lateral differentiation absent; L = 1–2.

I, ***Comesoma****, buccal cavity cup-shaped, no teeth; four cephalic setae and four subcephalic setae; lateral differentiation absent; spicules elongated; gubernaculum without apophysis; precloacal supplements present or absent; L = 1.5–5.

J, *Paracomesoma**, buccal cavity with three teeth; precloacal supplements present; L = 1.5–5.

K, *Metacomesoma*, buccal cavity without teeth; 10 cephalic setae; lateral differentiation absent; precloacal supplements absent; L = 1–2.

L, *Comesomoides*, 6 + 4 cephalic setae; lateral differentiation present; gubernaculum with apophysis; precloacal supplements present; L = 2.5–3.5.

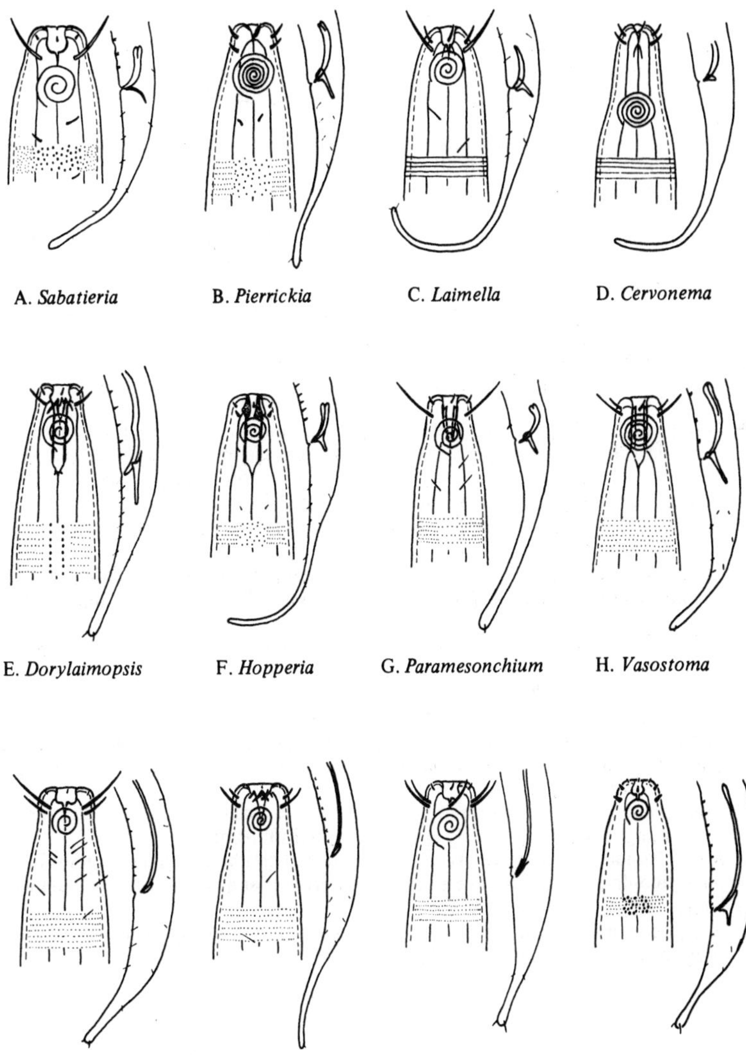
A. Sabatieria
B. Pierrickia
C. Laimella
D. Cervonema
E. Dorylaimopsis
F. Hopperia
G. Paramesonchium
H. Vasostoma
I. Comesoma
J. Paracomesoma
K. Metacomesoma
L. Comesomoides

Fig. 32.

A, ***Paracanthonchus****, buccal cavity with dorsal tooth, may be reduced; 10 cephalic setae; lateral differentiation of irregular dots; oesophageal bulb absent; tubular precloacal supplements; gubernaculum dentate distally, proximally paired; L = 1–4.

B, *Paracyatholaimus**, lateral differentiation absent; precloacal supplements setose or absent; gubernaculum distally simple, paired; L = 1–2.

C, *Acanthonchus*, precloacal supplements tubular, anterior-most larger; gubernaculum simple; L = 1–2.

D, *Paracyatholaimoides*, cephalic cuticle thick; male amphid larger than female amphid; tubular precloacal supplements; L = 0.5–1.5.

E, ***Cyatholaimus****, buccal cavity with dorsal tooth; 10 cephalic setae; lateral differentiation absent; oesophageal bulb absent; precloacal supplements absent; gubernaculum dentate distally, proximally single; L = 1–3.5.

F, *Praeacanthonchus**, teeth absent; lateral differentiation of larger dots; tubular precloacal supplements; gubernaculum dentate distally, proximally single; L = 1.5–3.

G, ***Marylynnia****, buccal cavity with dorsal tooth plus paired subventral teeth; 10 cephalic setae; lateral differentiation of larger and more widely spaced dots; oesophageal bulb absent; cup-shaped precloacal supplements; gubernaculum dentate distally, paired; tail elongated; L = 1.5–3.5.

H, *Longicyatholaimus*, similar to *Marylynnia* except: buccal cavity smaller; subventral teeth absent; L = 1.5–3.5.

I, *Paralongicyatholaimus**, teeth reduced or absent; lateral differentiation absent; oesophageal bulb present; precloacal supplements absent; gubernaculum simple; L = 0.5–3.

J, *Metacyatholaimus*, teeth reduced or absent; lateral differentiation of 3–5 longitudinal rows of dots; oesophageal bulb present; precloacal supplement absent; L = 0.5–1.5.

K, ***Pomponema****, buccal cavity with large dorsal tooth, two subventral teeth and denticles; 6 + 6 cephalic setae; amphid spiral; oesophageal bulb absent; cuticle punctated; lateral differentiation of longitudinal rows of dots; precloacal supplements consisting of several elements; L = 1–3.

L, *Minolaimus*, buccal cavity small, no teeth; 6 + 4 cephalic setae; lateral differentiation of larger and more widely spaced dots; tail filiform; L = 1–2.

M, *Nannolaimoides*, buccal cavity with small dorsal tooth; 10 cephalic setae; oesophageal bulb present; lateral differentiation present or absent; gubernaculum with lateral flange; L = 1.5–2.

N, *Kraspedonema*, buccal cavity with dorsal tooth; eight longitudinal files of dots; precloacal supplements simple; L = 1–2.

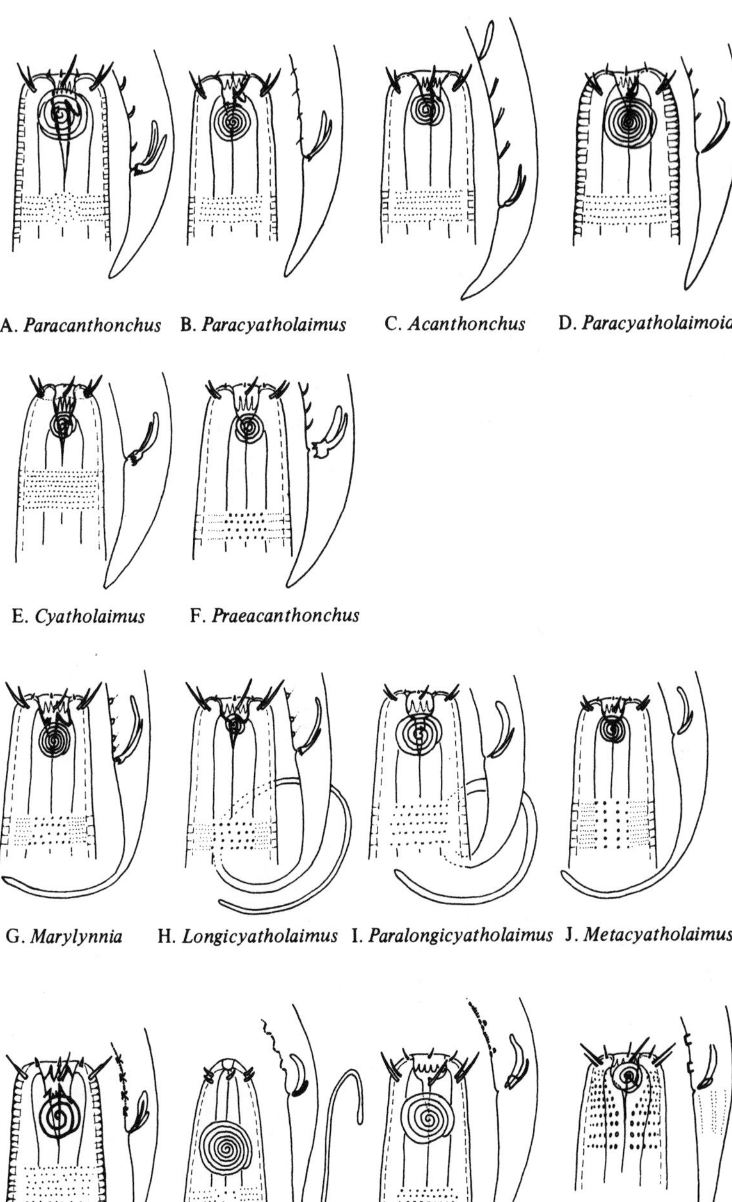
A. Paracanthonchus
B. Paracyatholaimus
C. Acanthonchus
D. Paracyatholaimoides
E. Cyatholaimus
F. Praeacanthonchus
G. Marylynnia
H. Longicyatholaimus
I. Paralongicyatholaimus
J. Metacyatholaimus
K. Pomponema
L. Minolaimus
M. Nannolaimoides
N. Kraspedonema

Fig. 33.

A, ***Neotonchus****, punctated cuticle with lateral differentiation (usually); sublateral cuticle pores present; buccal cavity with triangular dorsal tooth; 6 + 4 cephalic setae; amphid multi-spiral; oesophageal bulb present; 6–9 conspicuous cup-shaped supplements; tail conical; lateral row of punctations on posterior third of tail; L = 0.6–1.

B, *Neotonchoides**, similar to *Neotonchus* except: lateral differentiation present or absent; lateral cuticle pores; 8–14 precloacal supplements; lateral row of tail punctations absent; L = 0.6–1.5.

C, *Gomphionchus*, similar to *Neotonchoides* except: large forward-pointing dorsal tooth; 20 precloacal supplements; L = 0.9–1.

D, ***Gomphionema***, massive dorsal tooth; large posterior oesophageal bulb, one-third of oesophagus length; spicules and gubernaculum plus extra lateral pieces; 14–24 precloacal supplements; L = 0.8–1.5.

E, *Ethmolaimus*, large dorsal tooth plus two large subventral teeth; amphid single turn; tail conico-cylindrical; freshwater or brackish water habitat; L = 0.4–1.2.

F, *Trichethmolaimus*, extremely long somatic setae; elongated cuticle rods in head region; stout anterior cephalic setae; L = 0.8.

G, ***Nannolaimus***, minute buccal cavity; labial and anterior cephalic sensilla elongate; large multi-spiralled amphid; L = 1–1.5.

H, *Filitonchus*, labial sensilla minute; L = 1–1.5.

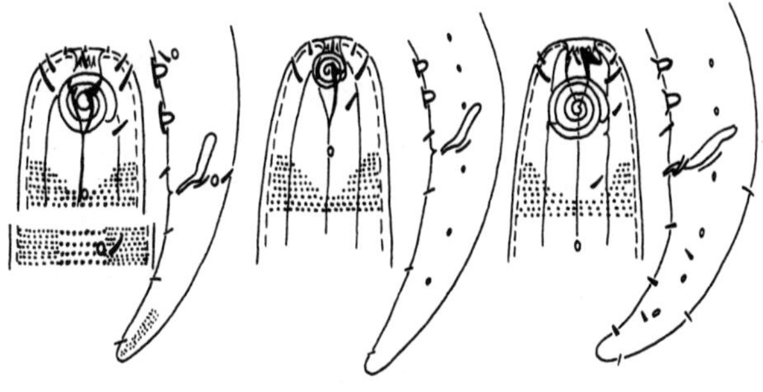

A. *Neotonchus* B. *Neotonchoides* C. *Gomphionchus*

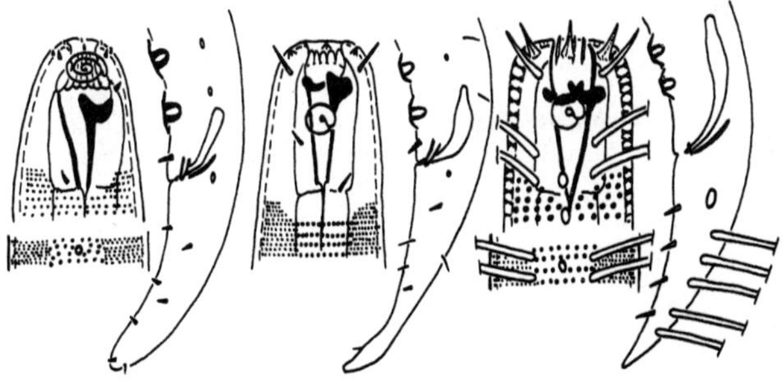

D. *Gomphionema* E. *Ethmolaimus* F. *Trichethmolaimus*

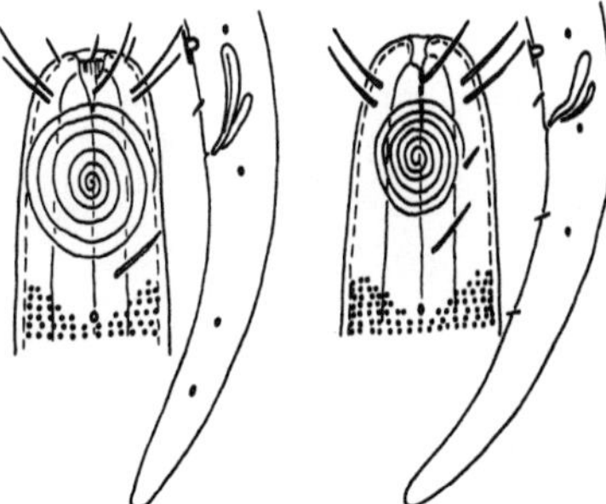

G. *Nannolaimus* H. *Filitonchus*

Fig. 34.

A, ***Synonchiella****, buccal cavity with three jaws, distally double; amphid spiral; 10 cephalic setae; lateral differentiation absent; oesophageal bulb absent; cup-shaped precloacal supplements; L = 1.5–2.5.

B, *Synonchium*, buccal cavity with three jaws, distally single; cephalic sensilla papilliform; lateral differentiation of wider spaced transverse rows of dots; precloacal supplements small or absent; gubernaculum absent; tail short; L = 1–2.

C, *Cheironchus*, dorsal jaw reduced; prominent pharyngeal bulb; oesophageal bulb present; lateral differentiation present; cup-shaped precloacal supplements; L = 1.5–3.

D, *Demonema*, jaws with knobbled distal end, dorsal jaw reduced; oesophageal bulb absent; setose precloacal supplements; tail long; L = 1–2.

E, ***Halichoanolaimus****, buccal cavity in two parts, separated by row of odontia; lateral differentiation absent; setose precloacal supplements; tail long; L = 0.5–2.5.

F, *Gammanema**, cup-shaped precloacal supplements; tail short; L = 1.5–4.5. *Choniolaimus**, similar to *Gammanema* except: oesophageal bulb present; L = 1.5–2.5.

G, *Latronema*, cuticle with 12–50 longitudinal rows of dots; 10 equally spaced cephalic setae; L = 1–2.

H, ***Richtersia****, 20–40 longitudinal rows of small spines; six cephalic setae + six subcephalic setae; buccal cavity unarmed; oesophageal bulb absent; L = 0.8.

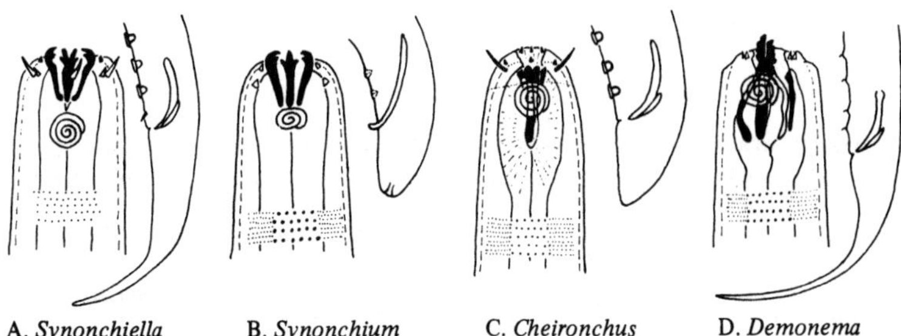

A. *Synonchiella* B. *Synonchium* C. *Cheironchus* D. *Demonema*

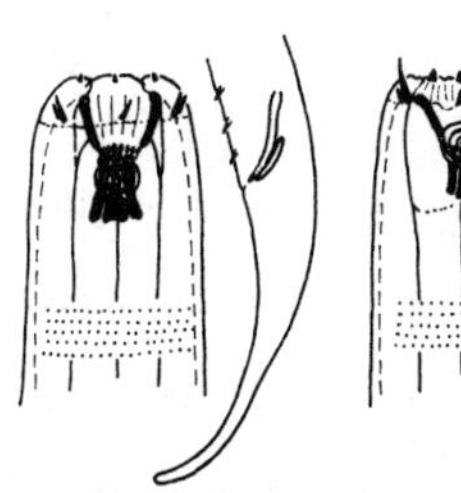

E. *Halichoanolaimus*

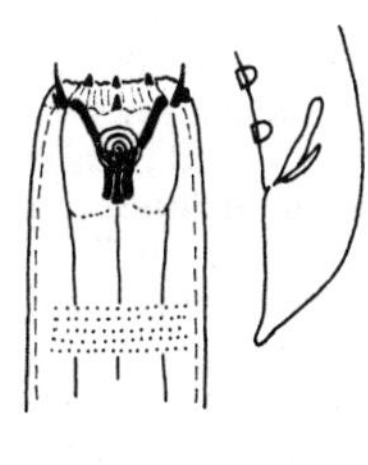

F. *Gammanema*

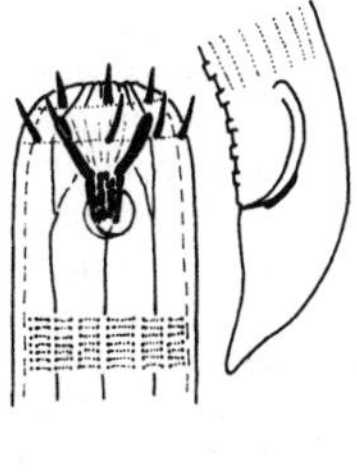

G. *Latronema*

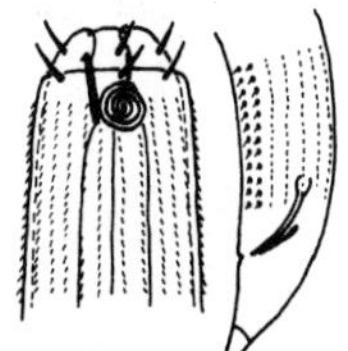

H. Richtersia

Fig. 35. Drawings beneath the heads show the cuticle pattern in the mid-body region.

A, ***Nygmatonchus***, amphid conspicuous; 6 + 4 cephalic setae; buccal cavity with dorsal tooth; cuticle heterogeneous with lateral differentiation; precloacal supplements absent; L = 0.5–1.5.

B, *Actinonema**, amphid conspicuous; cuticle heterogeneous with lateral differentiation beginning at the end of the oesophagus; L = 0.5–1.5.

C, *Rhips**, lateral differentiation begins at the end of the oesophagus; spicules double-jointed; spicules and gubernaculum plus lateral pieces; L = 1–1.5.

D, *Trochamus*, amphid indistinct; L = 0.5.

E, ***Neochromadora****, buccal cavity conical; one hollow dorsal tooth and two subventral teeth; four cephalic setae; cuticle heterogeneous with lateral differentiation of two or three longitudinal rows of dots; distinct posterior oesophageal bulb present; precloacal supplements cup-shaped; L = 0.5–2.

F, *Chromadorella**, buccal cavity with three solid teeth; cuticle heterogeneous with lateral differentiation of two to four longitudinal rows of dots; L = 0.5–1.5.

G, ***Chromadora****, buccal cavity conical; one dorsal and two subventral solid teeth; four cephalic setae; cuticle homogeneous with lateral differentiation of four longitudinal rows of dots; precloacal supplements present; L = 0.5–1.5.

H, *Dichromadora**, dorsal tooth hollow; cuticle homogeneous with lateral differentiation of two longitudinal rows of dots; L = 0.5–1.5.

I, *Spilophorella**, cuticle with lateral differentiation; oesophageal bulb double; precloacal supplements absent; tail tip elongated; L = 0.5–1.5.

J, *Atrochromadora**, teeth solid; amphid spiral; cuticle homogeneous with lateral differentiation; precloacal supplements present; L = 0.5–1.

K, ***Hypodontolaimus****, large S-shaped hollow dorsal tooth; four cephalic setae; cuticle homogeneous with lateral differentiation of 2–4 longitudinal rows of dots; precloacal supplements present (usually); L = 0.5–1.5.

L, *Ptycholaimellus**, similar to *Hypodontolaimus* except: oesophageal bulb double; L = 0.5–1.5.

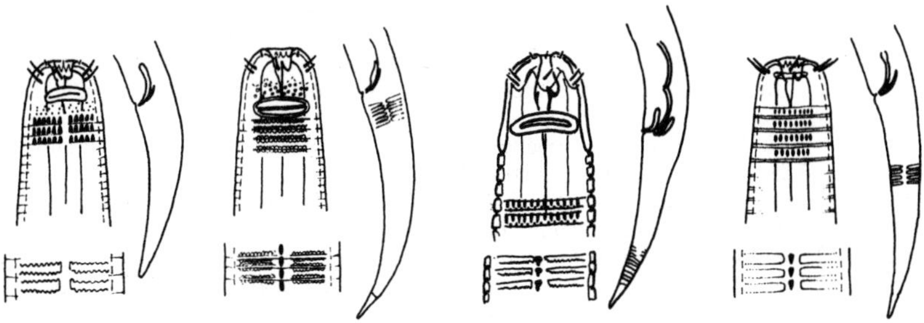

A. *Nygmatonchus* B. *Actinonema* C. *Rhips* D. *Trochamus*

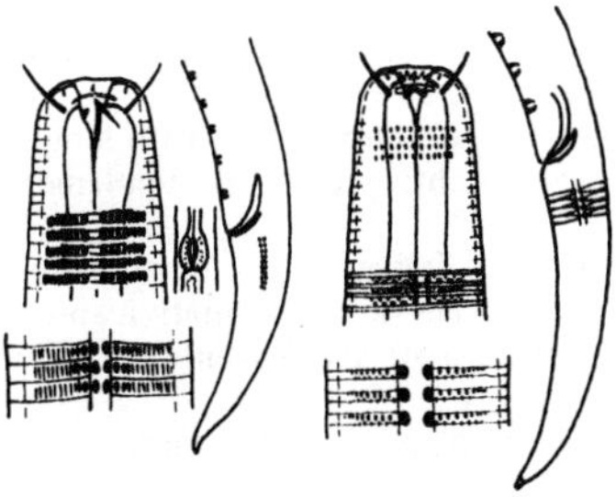

E. *Neochromadora* F. *Chromadorella*

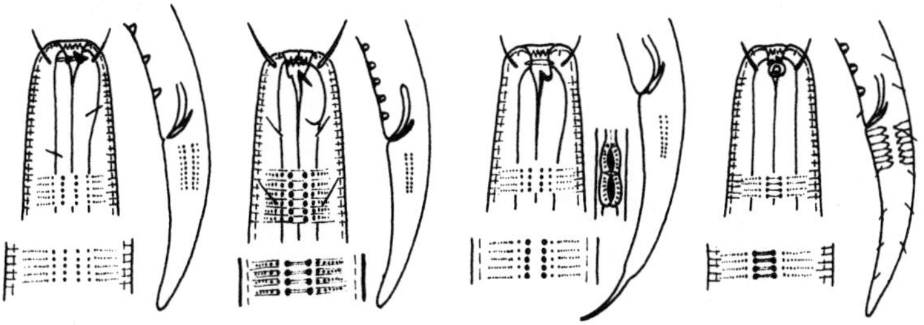

G. *Chromadora* H. *Dichromadora* I. *Spilophorella* J. *Atrochromadora*

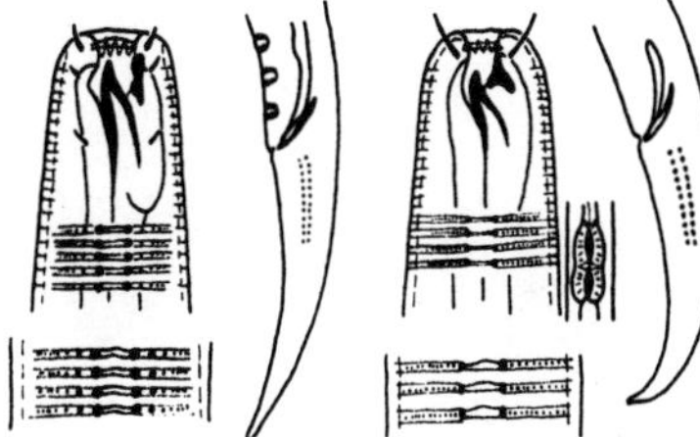

K. *Hypodontolaimus* L. *Ptycholaimellus*

Fig. 36.

A, ***Chromadorita********, dorsal tooth hollow; four cephalic setae; cuticle homogeneous; lateral rows absent although lateral dots are sometimes larger than the median ones; precloacal supplements present or absent; L = 0.5–1.5.

B, *Innocuonema**, cuticle heterogeneous, dots larger anteriorly; lateral differentiation absent; precloacal supplements absent; L = 0.5–1.

C, *Parachromadorita*, amphid oval, conspicuous; cuticle homogeneous, lateral dots enlarged; precloacal supplements present; L = 0.5–1.5.

D, ***Prochromadorella********, three solid teeth; four cephalic setae; cuticle heterogenous, no longitudinal rows but lateral dots enlarged; precloacal supplements present (usually); L = 0.5–1.5.

E, *Chromadorina**, cuticle homogeneous, lateral differentiation absent; precloacal supplements present or absent; L = 0.5–1.

F, *Prochromadora**, single solid dorsal tooth; cuticle homogeneous, lateral differentiation absent; precloacal supplements present; L = 0.5–2.5.

G, ***Spiliphera********, three solid teeth; amphid a distinct spiral; cuticle homogeneous, lateral differentiation absent; precloacal supplements absent; L = 0.5–1.5.

H, *Acantholaimus*, three or more teeth; amphid large, circular; cuticle with longitudinal rows present or absent; gubernaculum absent; precloacal supplements absent; L = 0.5–2.5.

I, ***Euchromadora********, buccal cavity with large dorsal tooth, subventral teeth and rows of denticles; four cephalic setae; cuticle heterogeneous, lateral differentiation absent; oesophageal bulb present; precloacal supplements absent; spicules and gubernaculum plus extra large L-shaped lateral pieces; L = 1–4.

J, *Graphonema**, similar to *Euchromadora* except: no denticles in the buccal cavity; L = 1–2.

K, *Parapinnanema**, similar to *Graphonema* except: ventral precloacal thickening in cuticle; L = 1–4.

L, *Steineridora**, similar to *Euchromadora* except: large rectangular dorsal tooth; oesophageal bulb present; L = 1–3.

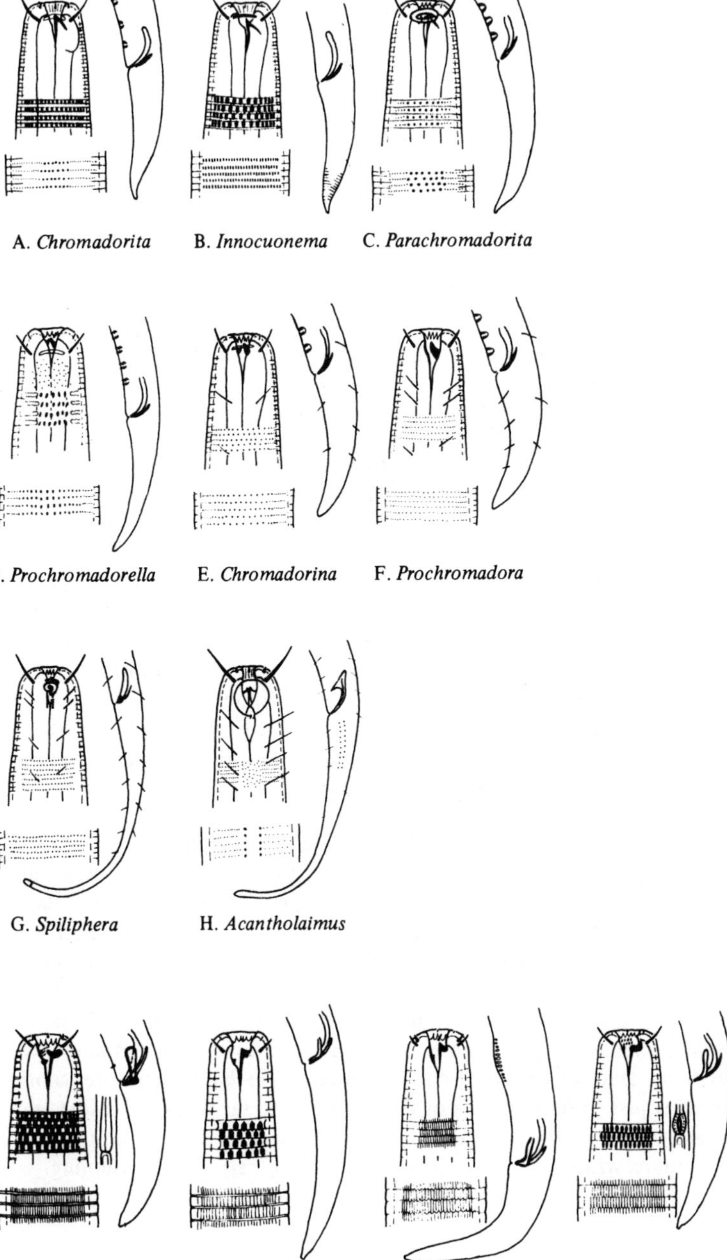

A. *Chromadorita* B. *Innocuonema* C. *Parachromadorita*

D. *Prochromadorella* E. *Chromadorina* F. *Prochromadora*

G. *Spiliphera* H. *Acantholaimus*

I. *Euchromadora* J. *Graphonema* K. *Parapinnanema* L. *Steineridora*

Introduction to Enoplia

The Subclass Enoplia includes the largest species of marine nematodes and for this reason members of this group are encountered frequently in general faunistic studies which are not specially concerned with the meiofauna. We will use 'enoplid' for a member of the Enoplia, as this term is widely used among nematologists.

The principal feature which distinguishes the Enoplia from the other major subclass of marine nematodes, the Chromadoria, is that the amphids (sense organs on the head) are not primarily spiral in structure; this usually means that they are pocket-like but they may also have other manifestations including the secondary spiral form found in *Bathylaimus* and *Tripyloides*, so that this feature is not always of much use when deciding whether the species under the microscope is an enoplid or not. The cuticle in the Enoplia is generally smooth, but may be marked with fine irregular punctations, as in some species of *Enoplus*, and in many genera fine transverse striations can be discerned, although the cuticle is never marked with large, regularly arranged punctations or coarse annulation. However, the cuticle is also smooth in some Chromadoria. The most important order of the Enoplia (in terms of numbers of species and genera) is the Enoplida, all species of which have specialised stretch receptors in the lateral fields called **metanemes** (see p. 87) which are not found in any other nematode group. However, species in the less important order, Trefusiida, lack metanemes. Thus, in practice, the best way to decide whether the species in question is an enoplid is whether or not it keys out to one of the enoplid genera in the pictorial guide.

In the general introduction we emphasised the paucity of marine nematode studies in the British Isles and the strong probability that genera previously unrecorded from our shores would be encountered; our solution there was to provide a key to the world genera. This problem is even more acute at the species level and almost any collection, especially from sublittoral habitats, is likely to contain species new to the British list. The provision of keys to the world species is not practicable, but keys to the British species alone would inevitably lead to many misidentifications. We have therefore confined ourselves just to descriptions of those species so far recorded. Most genera contain rather few species and the task of checking against all the species descriptions within the genus is not likely to be very time-consuming.

Structure

The external appearance and internal anatomy of marine nematodes in general have been described earlier (p. 2–10) and only the special features of the Enoplia will be dealt with here.

Head and buccal cavity

Inglis (1964) made a detailed study of the enoplid head and introduced some rather specialised terminology which will be largely followed in this volume. The terminology is best illustrated by reference to the structure of the head of a generalised member of the Family Thoracostomopsidae (Fig. 37).

Where the oesophagus is attached to the body wall in the head region the inner layer of body cuticle is modified to form the **cephalic capsule**, which is often thickened along its anterior margin to form the **cephalic ring**. The capsule is frequently divided into lobes posteriorly by six **incisions** which terminate in rounded **fenestrae** accommodating the bases of the cephalic setae when these arise from the region of the cephalic capsule. The remainder of the body anterior to the oesophagus, with the appearance of a fluid-filled space, is the **cephalic ventricle**. The mouth is surrounded by three **lip-lobes**, which may have a distinct anterior **subsidiary lobe**. The mouth contains three **mandibles** representing curved thickenings of the lining of the buccal cavity; this curvature can give the appearance in optical section of two lateral rods united by an anterior curved bar. Often a distinctly striated semi-circular zone of cuticle flanking the mandibles, the **semi-lunar striations**, can be seen on the inner surface of the lip. Posteriorly the mandibles flare out to form a transverse bar, the **mandibular ring**. Three teeth or **onchia** are situated behind the mandibles, arising from the **onchial plate** which is intimately fused to the mandibles. The onchial plate is supported by **radial processes** which, in conjunction with the mandibular ring, are linked to the cephalic ring by **radial masses**.

All these structures are, of course, not found throughout the Enoplia. For example, in many genera the oesophagus does not fuse anteriorly with the body wall, so there is no cephalic capsule, and many lack the mandibles, onchia and associated structures. The complex head structure in the Thoracostomopsidae simply serves to illustrate the terminology.

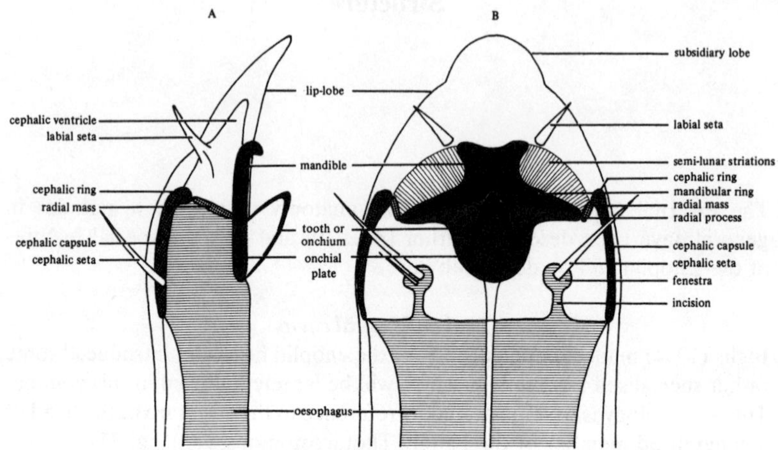

Fig. 37. Structure of the head in the Thoracostomopsidae; A, longitudinal section through one lip; B, plan.

Metanemes

Metanemes are filamentous organs which have been found in all 16 families of the Order Enoplida, but in no other nematodes (Lorenzen, 1978b; 1981b). They are arranged serially in the epidermal cords, either parallel to (**orthometanemes**) or at an angle of 10–30° to the longitudinal axis of the body (**loxometanemes**). Each consists of a 5–15 μm long **scapulus**, a 20–400 μm long cuticularised **frontal filament** and in most taxa a 30–170 μm **caudal filament** (Fig. 38). A **sensory cell**, which may be difficult to see, surrounds the scapulus. These organs are considered to be stretch receptors which play a role in co-ordinating body movements and controlling body volume.

Their number, form and arrangement varies considerably, and for most of the species described in this synopsis the arrangement is not known. We have not included descriptions of these organs even in species where they have been described, because at present they are of little use for species identification, although their presence is certain confirmation that a species is in the Enoplida.

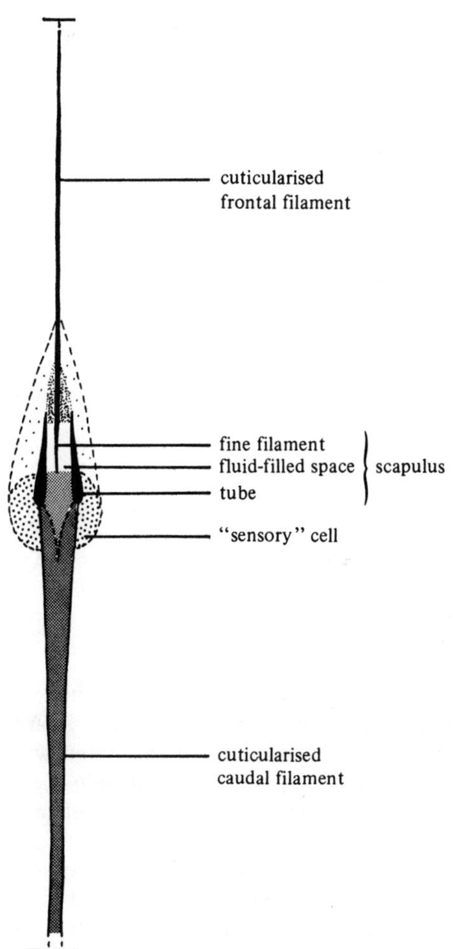

Fig. 38. Diagram of a metaneme (from Lorenzen, 1978b).

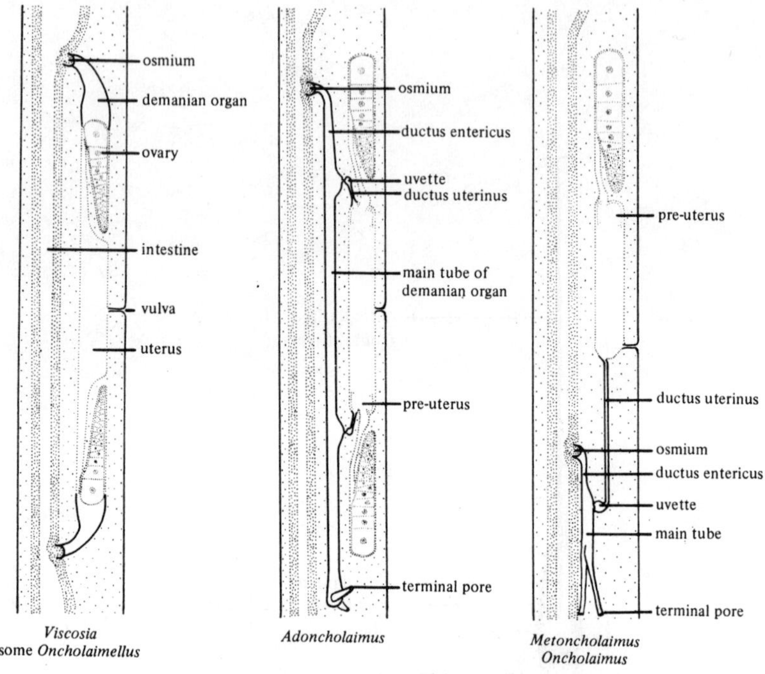

Fig. 39. Diagrammatic representation of the arrangement of the demanian system in oncholaimids (from Rachor, 1969).

Demanian system

The **demanian system** is a more or less complex seminal receptacle found in the females of most genera in the Oncholaimidae, a large and important enoplid family. A comprehensive account of this system was given by Rachor (1969). The spermatozoa are kept alive with the help of secretions from the **osmium**, which consists of modified intestinal epithelium connected to the demanian system.

Of the genera described in this synopsis, *Pontonema* lacks a demanian system. The simplest system is found in *Viscosia* and some *Oncholaimellus* species in which the demanian system is a pocket off each ovary, from the latter's point of flexure. The organ unites with the intestinal epithelium (Fig. 39). More complex systems comprise a longitudinal tube between the body wall and intestine, with an osmium anteriorly and **terminal pores**, varying in number and position, opening posteriorly to the exterior. The main tube is connected to the uterus by narrow ducts, each called a **ductus uterinus**; two in the didelphic genus *Adoncholaimus* and one in the monodelphic genera *Oncholaimus*, *Metoncholaimus* and *Metaparoncholaimus* (Fig. 39). The junction of the main tube and ductus uterinus is swollen into a structure called the **uvette**, which is sometimes surrounded by star-like **moniliform glands**. The section of the main tube between the osmium and uvette is called the **ductus entericus**.

Features of the demanian system are often difficult to see, but the position and number of the terminal pores (e.g. Fig. 93) and the position and structure of the uvette (e.g. Fig. 95) are often prominent features of value in determining species. In general, however, we have not described this system in cases where it is of little taxonomic use.

Biology

The enoplids have a larger proportion of species with strong jaws, presumed to be predators, than either the chromadorids or monhysterids, although rather little is known about their actual feeding behaviour. Several members of the Thoracostomopsidae from preserved samples have been found in the process of ingesting other nematodes, and representatives of the Oncholaimidae will feed on nematodes in the laboratory. The oncholaimid *Adoncholaimus thalassophygas* produces mucous agglutinations; the juveniles are thought to feed primarily on dissolved organic matter released by intense microbial activity within the mucus but do not utilise micro-organisms directly. The adults, although supplementing their diet by predation and scavenging, also retain the ability to absorb dissolved organics (Lopez *et al.*, 1979). On the other hand, the juveniles of another oncholaimid, *Oncholaimus oxyuris*, feed on bacteria and algae in the laboratory (Heip *et al.*, 1978). Meyers and Hopper (1966) found that fungal-degraded cellulose substrates attracted large numbers of gravid females of the oncholaimid *Metoncholaimus scissus*, and egg-clusters were laid in the substrate (Hopper and Meyers, 1966). Lopez *et al.* (1979) regard this as an example of brood care, the newly hatched juveniles being placed in an environment where the concentration of dissolved organic matter is high.

Some of the larger enoplids have relatively long generation times. *Enoplus communis* on the eastern coast of the USA has an annual life cycle in the field (Wieser and Kanwisher, 1960). Among the oncholaimids, field data supported by experiments in which the animals have been reared in the laboratory indicate that *Oncholaimus oxyuris* has either two generations per year or three generations in two years (Smol *et al.*, 1980; Heip *et al.*, 1978) and that *Oncholaimus brachycercus* has an annual life cycle, or two generations in warm years (Skoolmun and Gerlach, 1971; Gerlach and Schrage, 1972). Many species, however, have shorter generation times (Hopper *et al.*, 1973). Little is known of the reproductive behaviour of enoplids because, apart from some oncholaimids, they are difficult to rear in the laboratory. *Enoplus communis* produces egg-strings in an adhesive sheath (Hopper, 1961), but this does not appear to be common in marine nematodes, although some may lay their eggs in an adhesive mass.

Classification

The divisions of the Subclass Enoplia to family level, based on the recent revision by Lorenzen (1981a), are given in Fig. 40. The following list of families, genera and species only includes those recorded from the British Isles.

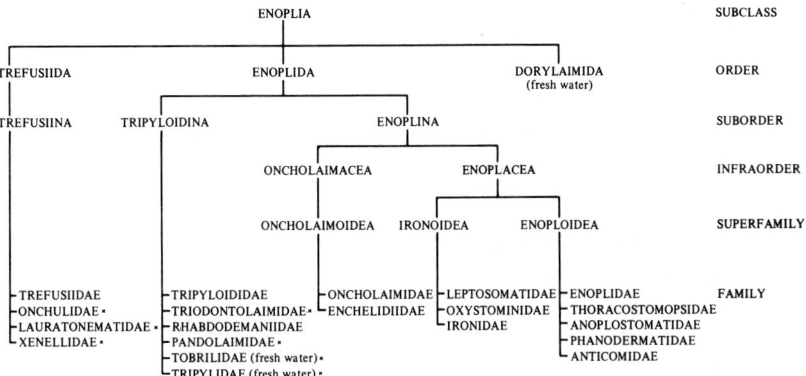

Fig. 40. Classification of the Enoplia to family level, based on Lorenzen (1981a). Families marked with a cross are not included in this Synopsis.

Family ENOPLIDAE
  Enoplus Dujardin, 1845
    E. brevis Bastian, 1865
    E. paralittoralis Wieser, 1953
    E. communis Bastian, 1865
    E. quadridentatus Berlin, 1853
    E. schulzi Gerlach, 1952

Family THORACOSTOMOPSIDAE
  Enoploides Ssaweljev, 1912
    E. brunettii Gerlach, 1953
    E. longispiculosus Vitiello, 1967
    E. spiculohamatus Schulz, 1932
    E. labrostriatus (Southern, 1914)
  Enoplolaimus De Man, 1893
    E. propinquus De Man, 1922
    E. denticulatus Warwick, 1970
    E. litoralis Schulz, 1936
    E. longicaudatus (Southern, 1914)
    E. vulgaris De Man, 1893
    E. subterraneus Gerlach, 1952
  Epacanthion Wieser, 1953
    E. gorgonocephalum Warwick, 1970
    E. buetschlii (Southern, 1914)
    E. mawsoni Warwick, 1977
  Mesacanthion Filipjev, 1927
    M. africanthiforme Warwick, 1970
    M. diplechma (Southern, 1914)
    M. hirsutum Gerlach, 1952
  Paramesacanthion Wieser, 1953
    P. hirsutum Warwick, 1970
    P. marei Warwick, 1970
  Thoracostomopsis Ditlevsen, 1918
    T. doveae Warwick, 1970
  Trileptium Cobb, 1933
    T. parisetum Warwick and Platt, 1973

Family ANOPLOSTOMATIDAE
Anoplostoma Bütschli, 1874
A. viviparum (Bastian, 1865)
Chaetonema Filipjev, 1927
C. riemanni Platt, 1973

Family PHANODERMATIDAE
Crenopharynx Filipjev, 1934
C. marioni (Southern, 1914)
Phanoderma Bastian, 1865
P. albidum Bastian, 1865
P. cocksi Bastian, 1865
P. laticolle (Marion, 1870)

Family ANTICOMIDAE
Anticoma Bastian, 1865
A. acuminata (Eberth, 1863)
A. eberthi Bastian, 1865

Family IRONIDAE
Dolicholaimus De Man, 1888
D. marioni De Man, 1888
Thalassironus De Man, 1889
T. britannicus De Man, 1889
Trissonchulus Cobb, 1920
T. benepapillosus (Schulz, 1935)
T. obtusus (Bresslau and Stekhoven, 1935)
T. oceanus Cobb, 1920

Family LEPTOSOMATIDAE
Cylicolaimus De Man, 1889
C. magnus (Villot, 1875)
Leptosomatides Filipjev, 1918
L. euxinus Filipjev, 1918
Leptosomatum Bastian, 1865
L. bacillatum (Eberth, 1863)
L. elongatum Bastian, 1865
Platycoma Cobb, 1894
P. cephalata Cobb, 1894
Pseudocella Filipjev, 1927
P. coecum (Ssaweljev, 1912)
P. trichodes (Leuckart, 1849)
Synonchus Cobb, 1894
S. fasciculatus Cobb, 1894
S. brevisetosus (Southern, 1914)
S. longisetosus (Southern, 1914)
Thoracostoma Marion, 1870
T. coronatum (Eberth, 1863)

Family OXYSTOMINIDAE

Halalaimus De Man, 1888

H. gracilis De Man, 1888

H. longicaudatus (Filipjev, 1927)

H. isaitshikovi (Filipjev, 1927)

H. capitulatus Boucher, 1977

H. leptosoma (Southern, 1914)

H. longicollis Allgén, 1932

Oxystomina Filipjev, 1921

O. elongata (Bütschli, 1874)

O. asetosa (Southern, 1914)

O. unguiculata Stekhoven, 1935

Nemanema Cobb, 1920

N. cylindraticaudatum (De Man, 1922)

Paroxystomina Micoletzky, 1924

P. asymmetrica Micoletzky, 1924

Thalassoalaimus De Man, 1893

T. tardus De Man, 1893

Family ONCHOLAIMIDAE

Adoncholaimus Filipjev, 1918
- *A. fuscus* (Bastian, 1865)
- *A. thalassophygas* (De Man, 1876)
- *A. panicus* Cobb, 1930

Metaparoncholaimus De Coninck and Stekhoven, 1933
- *M. campylocercus* (De Man, 1876)

Metoncholaimus Filipjev, 1918
- *M. albidus* (Bastian, 1865)
- *M. scanicus* (Allgén, 1935)

Oncholaimellus De Man, 1890
- *O. calvadosicus* De Man, 1890
- *O. mediterraneus* Stekhoven, 1942

Oncholaimus Dujardin, 1845
- *O. brachycercus* De Man, 1889
- *O. dujardinii* De Man, 1876
- *O. oxyuris* Ditlevsen, 1911
- *O. campylocercoides* De Coninck and Stekhoven, 1933
- *O. skawensis* Ditlevsen, 1921
- *O. attenuatus* Dujardin, 1845

Pontonema Leidy, 1855
- *P. vulgare* (Bastian, 1865)
- *P. macrolaimus* (Southern, 1914)
- *P. simile* (Southern, 1914)

Viscosia De Man, 1890
- *V. viscosa* (Bastian, 1865)
- *V. abyssorum* (Allgén, 1933)
- *V. cobbi* Filipjev, 1918
- *V. glabra* (Bastian, 1865)
- *V. langrunensis* (De Man, 1890)
- *V. elegans* (Kreis, 1924)

Family ENCHELIDIIDAE

Belbolla Andrássy, 1973

B. gallanachmorae (Inglis, 1961)

Calyptronema Marion, 1870

C. maxweberi (De Man, 1922)

Eurystomina Filipjev, 1921

E. ornata (Eberth, 1863)

E. cassiterides (Warwick, 1977)

E. terricola (De Man, 1907)

Pareurystomina Mikoletzky, 1930

P. scilloniensis Warwick, 1977

P. acuminata (De Man, 1889)

Symplocostoma Bastian, 1865

S. tenuicolle (Eberth, 1863)

Family TRIPYLOIDIDAE

Bathylaimus Cobb, 1894

B. australis Cobb, 1894

B. capacosus Hopper, 1962

B. inermis (Ditlevsen, 1918)

B. paralongisetosus Stekhoven and De Coninck, 1933

B. stenolaimus Stekhoven and De Coninck, 1933

B. tenuicaudatus (Allgén, 1933)

Gairleanema Warwick and Platt, 1973

G. anagremilae Warwick and Platt, 1973

Tripyloides De Man, 1886

T. gracilis (Ditlevsen, 1918)

T. marinus (Bütschli, 1874)

Family RHABDODEMANIIDAE

Rhabdodemania Baylis and Daubney, 1926

R. major (Southern, 1914)

R. minor (Southern, 1914)

R. imer Warwick and Platt, 1973

Family TREFUSIIDAE

Rhabdocoma Cobb, 1920

R. riemanni Jayasree and Warwick, 1977

Trefusia De Man, 1893

T. longicaudata De Man 1893

T. zostericola Allgén, 1933

Two species of *Rhabditis*, a genus of predominantly terrestrial species, are also included in this volume. They belong to the Class Secernentea, whereas all other marine genera belong to the Adenophorea.

Family RHABDITIDAE
Rhabditis Dujardin, 1845
R. marina Bastian, 1865
R. ehrenbaumi Bresslau and Stekhoven, 1935

Species descriptions

In the descriptions which follow we have used several abbreviations which were described more fully earlier: h.d. = head diameter; c.d. = corresponding (body) diameter; a.b.d. = anal or cloacal body diameter; a is the De Manian ratio of body length/maximum breadth. The size of each genus is indicated by giving the number of valid species so far recorded worldwide. Where important keys to genera are available they have been referenced, but it should be borne in mind that many of these may be out of date. Those wishing to follow up these keys will find a full reference to their source in 'The Bremerhaven Checklist of Aquatic Nematodes', Part 2 (Gerlach and Riemann, 1974), but we have included full references after 1972. The full list of synonyms for each species can also be found in the Checklist. Wherever possible we have made original drawings and descriptions from British material, but where this has not been possible the source has been acknowledged in the figure legend; again the full reference can be found in the Bremerhaven Checklist.

The recorded distribution of each species, as with any other underworked group of organisms, is more a reflection of where collections have been made than of the actual geographical distribution. Thus, a heavy bias towards Ireland and South West England results from the authors' activities in these areas. However, the number of records gives some indication of commonness or rarity. Frequently a species is confined to one type of habitat throughout its range, but this is by no means always the case.

Most of the descriptions are based on a rather small number of specimens, often collected at the same locality. Little is known about variations in body-size which might occur from place to place. Thus, when making identifications, more emphasis should be placed on the detailed morphology of the animal, as illustrated in the figures, than on absolute or relative measurements.

Genus ENOPLUS Dujardin, 1845

Key: Wieser, 1953
The genus *Enoplus* is easily distinguished by its low lips with papilliform labial sense organs, and solid mandibles with no onchia. The British species are relatively distinctive, *E. communis* being very common on rocky shores among seaweeds and *E. brevis* similarly frequent in the sediments. Specimens thought to be *E. communis* should be carefully checked against the drawing and description as there are several similar species with eyespots, trumpet-shaped supplement and spicules with scales which, although not yet recorded from Britain, may well appear in the British nematode fauna.
Species: 33.

Enoplus brevis Bastian, 1865
(Fig. 41)

Description. Body length 6.2–7.7 mm. Maximum diameter 130–170 μm (a = 37–55). Cuticle marked with small rounded punctations. Three low lips with six small labial papillae at their bases. Six long cephalic setae about 0.47 h.d., four shorter cephalic setae about two-thirds this length. A group of three short setae positioned laterally just posterior to the cephalic capsule. Mandibles solid, bilobed anteriorly, no onchia. No definite eyespots but some irregular pigmented areas present anteriorly. Amphids small rounded pockets with ovoid openings, 0.09 c.d. wide, just in front of posterior border of cephalic capsule. Oesophagus cylindrical, 0.13 times body length. Excretory pore one-fifth of way down oesophagus length, nerve ring 42% of way down oesophagus length. Body setae short and scattered. Tail about 2 a.b.d.

Male with two files of longish setae between supplement and cloaca, with a pair of short stout setae on the posterior lip of the cloaca. Gubernaculum complicated; roughly triangular with a rounded distal tip; two lateral projections on each side, a distal slender projection with a pair of teeth and a more proximal rounded projection. Spicules smooth, equal, arcuate, 1.5 a.b.d. (158–170 μm). Supplement trumpet-shaped, about 2 a.b.d. in front of cloaca, 69–85 μm long, rounded distal tip with a series of small teeth just proximal to it.

Ovaries paired, equal, opposed, reflexed. Vulva at about 55% of body length.

Distribution. A common species in the muddy intertidal or shallow sublittoral recorded from several localities around the British Isles, often in areas of reduced salinity.

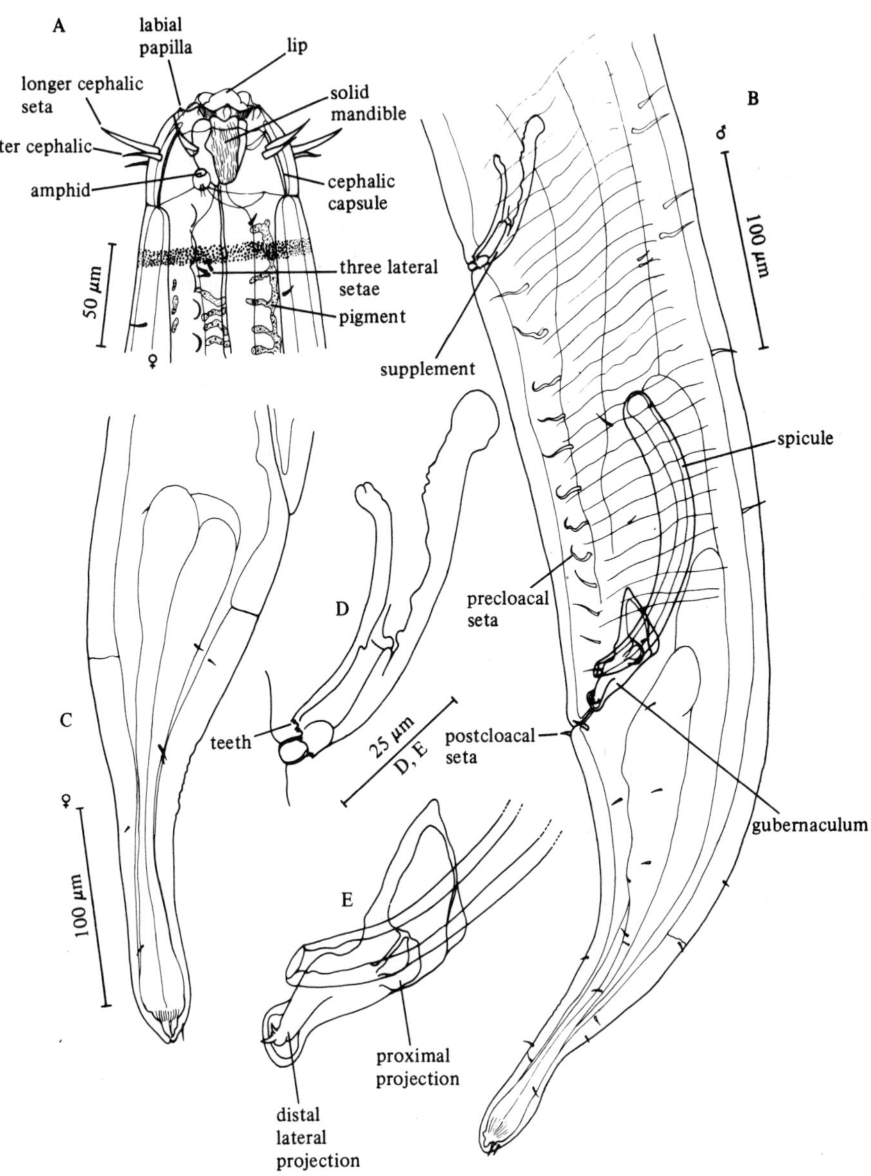

Fig. 41. *Enoplus brevis*. A, Female head; B, Male tail; C, Female tail; D, supplement; E, Gubernaculum and spicule tip. Original.

Enoplus paralittoralis Wieser, 1953
(Fig. 42)

Description. Body length 3.4 mm. Maximum diameter 84 μm (a = 40). Six longer cephalic setae 0.27 h.d., four others only slightly shorter. No eyespots, but irregular pigmented areas present. Amphids 0.08 c.d., at posterior margin of cephalic capsule. Tail conical, 1.6 a.b.d.

Gubernaculum a complex tubular structure with fenestrated walls about one-third of the length of the spicules. Spicules smooth, 1.3 a.b.d. (94 μm). Supplement tubular, bent, complex, 35 μm long, 2 a.b.d. in front of cloaca.

Distribution. A single male from the Exe estuary (high on the shore in muddy sand); two males from Isles of Scilly (rotting seaweed on strandline).

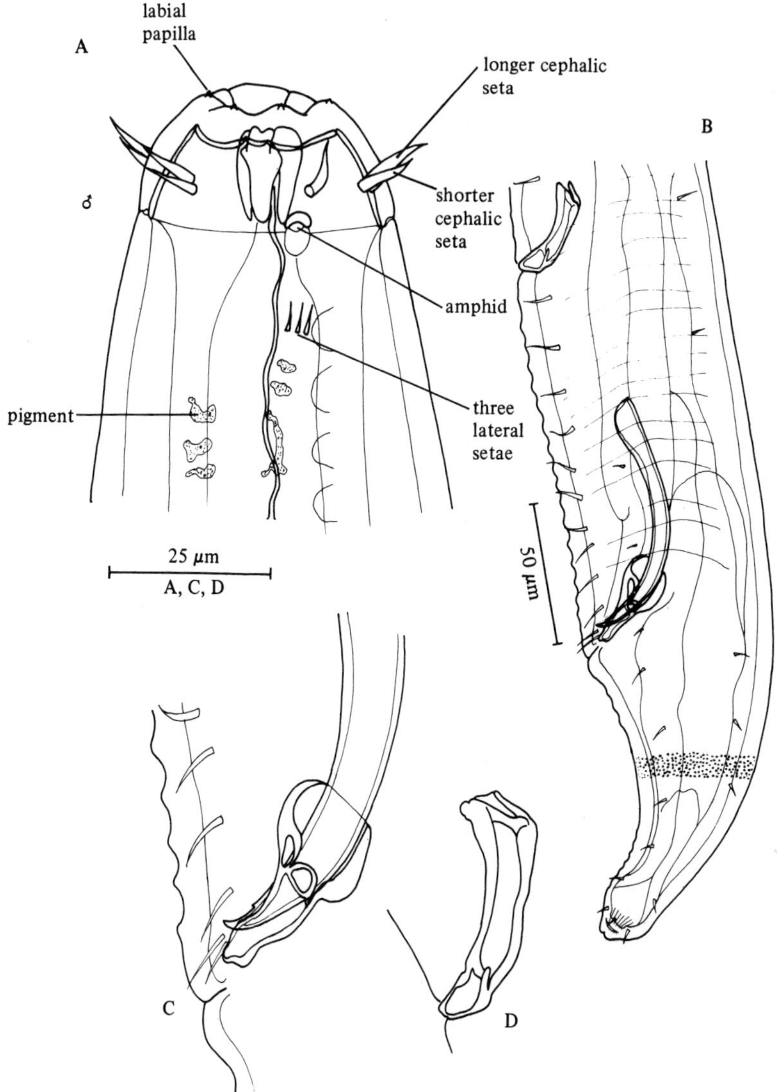

Fig. 42. *Enoplus paralittoralis*. A, Male head; B, Male tail; C, Gubernaculum and spicule tip; D, Supplement. Original.

Enoplus communis Bastian, 1865
(Fig. 43)

Description. Body length 5–10 mm (a = 25–50). Six longer cephalic setae 0.4 h.d., four others about two-thirds this length. Prominent pigmented eyespots. Amphids 0.1 c.d. wide, anterior to base of cephalic capsule. Tail a little over 2 a.b.d., posterior third cylindrical.

Gubernaculum in three pieces: paired lateral pieces with lateral wings; median ventral piece less obvious but composed of beard-like strands, best seen in ventral view. Spicules curved, 230 μm (1.4 a.b.d.), with four or five scale-like plates on their ventral surfaces near the middle. Precloacal supplement very large, three-quarters of the spicule length, trumpet shaped, about 1 tail length in front of cloaca.

Distribution. A very common species recorded from a large number of localities around the British Isles, typically found in seaweeds and holdfasts intertidally or in shallow water.

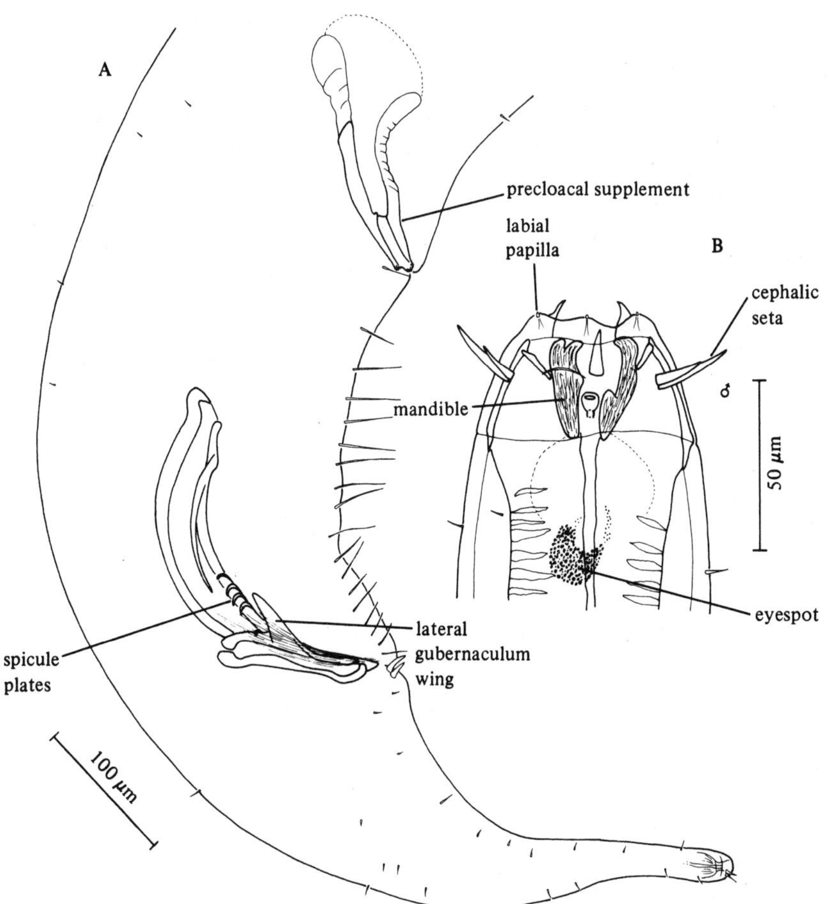

Fig. 43. *Enoplus communis*. A, Male tail; B, Male head. Original.

Enoplus quadridentatus Berlin, 1853
(Fig. 44)

Description. Length 3.1–4.8 mm. Maximum diameter 110–60 μm (a = 23–35). Very similar to, but smaller than, *E. communis*, distinguished from it by the relatively longer tail (a little over 3 a.b.d.), spicules 140 μm (1.7 a.b.d.), supplement only half the spicule length and differently shaped, with a characteristic tube lining the distal third. There are only two spicular scales on the specimen illustrated (from the Isles of Scilly), but up to eight have been recorded from Mediterranean specimens.

Distribution. Isles of Scilly (intertidal seaweeds). Undoubtedly rarer than *E. communis*, but may have been confused with it in the past.

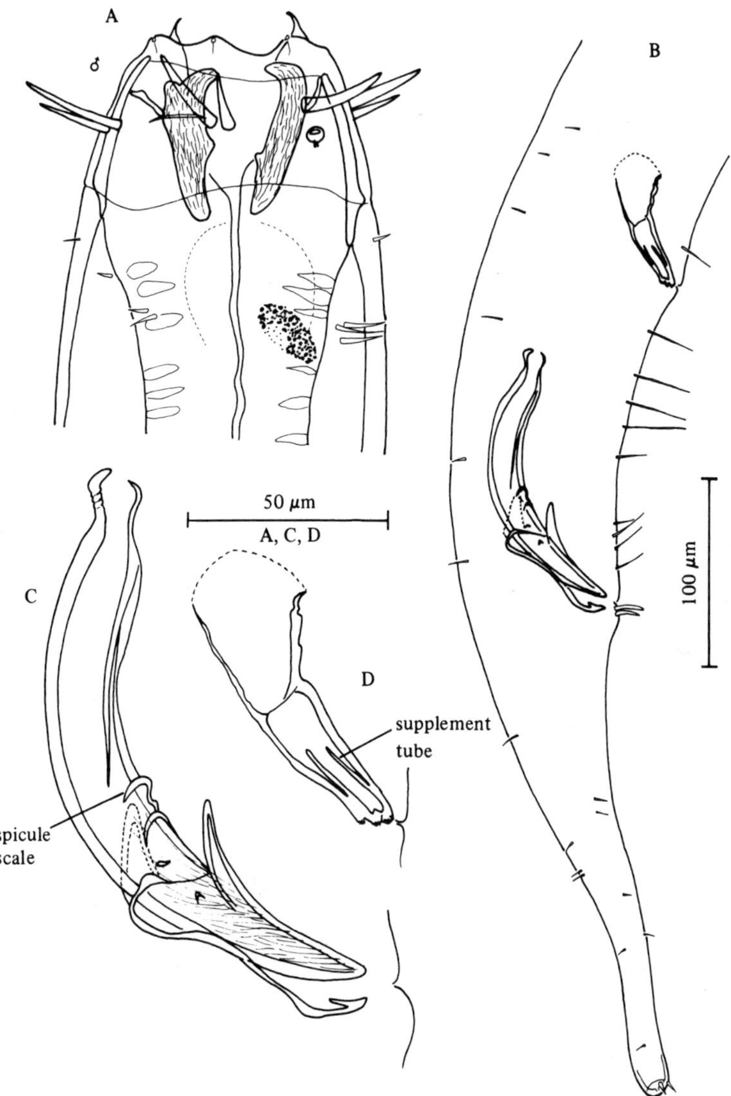

Fig. 44. *Enoplus quadridentatus*. A, Male head; B, Male tail; C, Spicules and gubernaculum; D, Supplement. Original.

Enoplus schulzi Gerlach, 1952
(Fig. 45)

Description. Body length 1.3–1.6 mm. Maximum diameter 45–56 μm (a = 23–32). Cuticle smooth. Six longer cephalic setae three-fifths of h.d. long, remaining four only slightly shorter. Amphids large, almost one-third of h.d. wide, situated on posterior border of cephalic capsule. Mandibles expanded in the middle but not bilobed anteriorly. Tail conical and pointed 1–1.6 a.b.d.

No rows of setae between supplement and cloaca in male. Gubernaculum simple, small, paired. Spicules smooth, simple, 62–63 μm (1.9 a.b.d.).

Supplement L-shaped, 22 μm, level with proximal ends of spicules, distal edge serrated.

Distribution. Exe estuary (gravelly sand at High Water Spring Tide level).

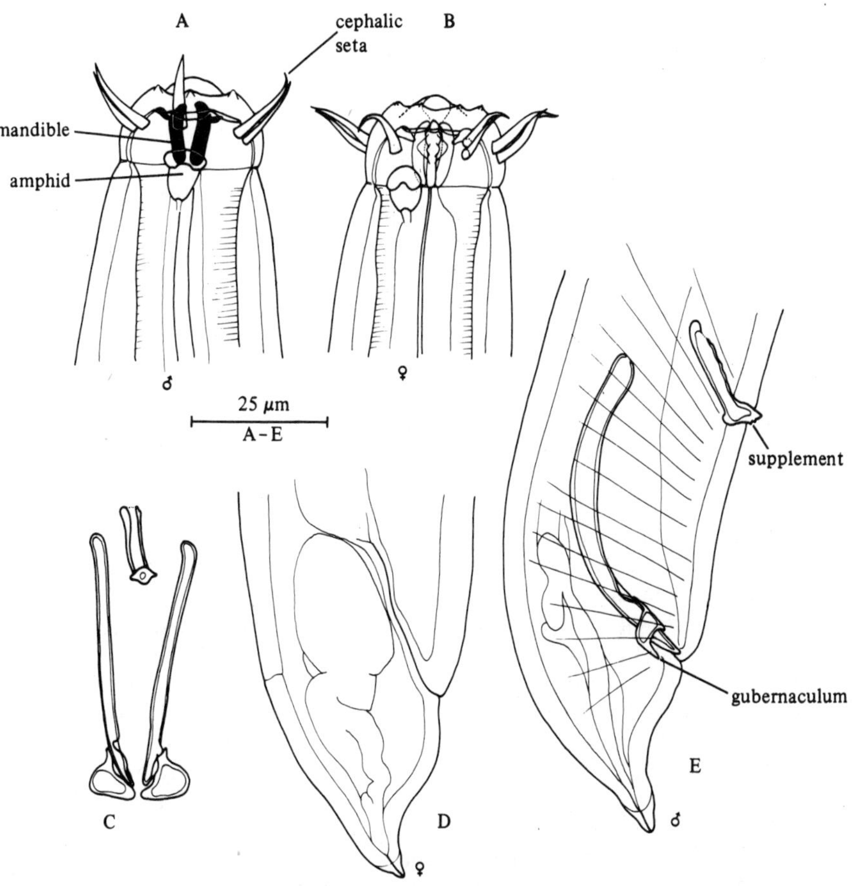

Fig. 45. *Enoplus schulzi*. A, Male head; B, Female head; C, Ventral view of spicules, gubernaculum and supplement; D, Female tail; E, Male tail. Original.

Genus ENOPLOIDES Ssaweljev, 1912

Keys: Wieser, 1953; Wieser and Hopper, 1967

Enoploides species are characterised by the solid bifurcate mandibles. Species are rather similar, and also somewhat variable in the lengths of cephalic setae and the tail. In the Exe estuary, specimens of *E. brunettii* from exposed beaches have more elongate cephalic setae and tails than those from more sheltered situations. Species are best separated on the detailed morphology of the male copulatory apparatus.

Species : 24

Enoploides brunettii Gerlach, 1953
(Fig. 46)

Description. Body length 2.5–3.9 mm. Maximum diameter 69–75 μm (a = 33–53). Cuticle smooth. Lips high, internal surfaces coarsely striated anterior to a curved line joining the tips of the mandibles and the lip bases, with fine semi-lunar striations posterior to this line. Outer margin of lip scalloped, except for subsidiary lobe (which has no internal striation). Labial setae 0.27–0.48 h.d. Cephalic setae at level of anterior margin of cephalic capsule; longer six 0.97–1.59 h.d., four shorter submedian setae 0.32–0.68 h.d. Six short subcephalic setae at level of bases of onchial plates. Three solid mandibles with bifurcate tips. Onchia small, equal in size. Triangular radial processes extend sideways from the onchia and two more cuticular thickenings extend backwards from each onchium. Crescentic radial masses are present at the level of the cephalic setae in the interlabial positions. Oesophagus with a rounded swelling anteriorly, 0.22 times body length. Tail 2.5–4.2 a.b.d. long, anterior half to two-thirds conical, remainder cylindrical.

Spicules elongate, 298–343 μm (5.75 a.b.d.), transversely striated.

Gubernaculum complex: paired tubular structures each swollen distally, terminating in a blunt tooth and a sharper lateral tooth, 47–62 μm long. Supplement simple, tubular, about half the spicule length in front of the cloaca. Male tail with a pair of prominent papillae just posterior to cloaca.

Ovaries paired, opposed, reflexed. Vulva at 56–59% of body length.

Distribution. Exe estuary; Isles of Scilly (intertidal sand).

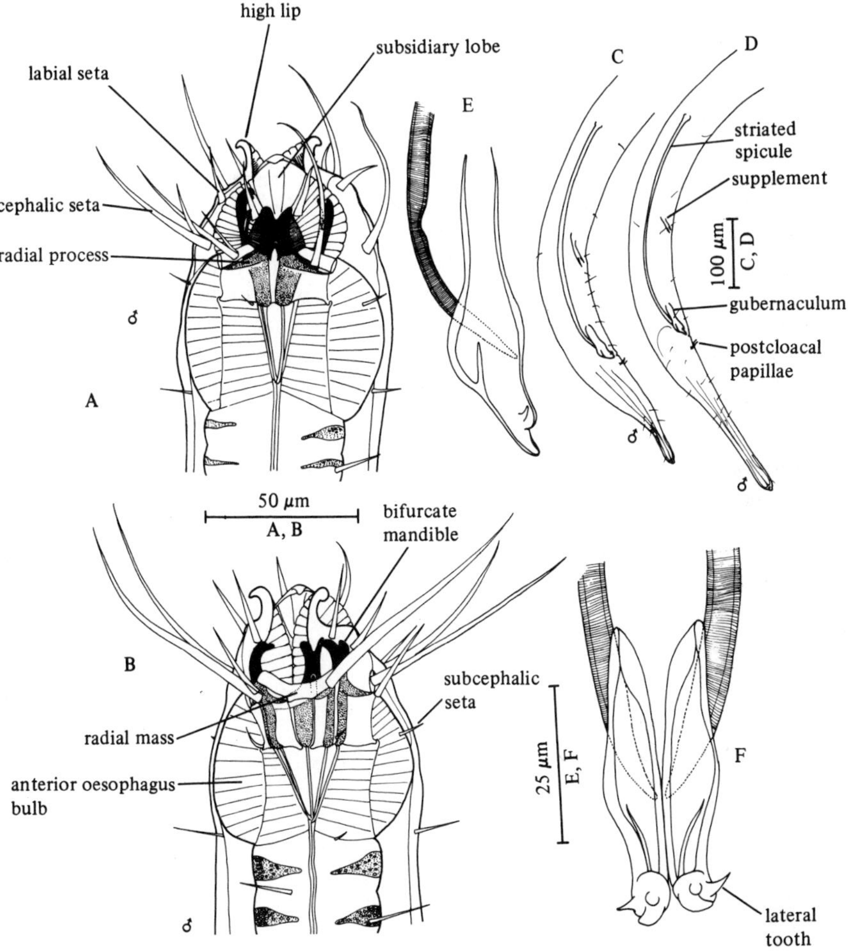

Fig. 46. *Enoploides brunettii*. A, Male head (from sheltered beach); B, Male head (from exposed beach); C, D, Range in form of male tail; E, Gubernaculum and spicule tip (lateral); F, Gubernaculum and spicule tips (ventral). Original.

Enoploides longispiculosus Vitiello, 1967
(Fig. 47)

Description. Body length 2.4–2.9 mm. Maximum diameter 92–112 μm (a = 25–30). Shorter and stouter than *E. brunettii.* Cuticle smooth. Lips as in *E. brunettii.* Labial setae 0.22–0.29 h.d. Six longer cephalic setae 0.8–0.96 h.d., shorter four 0.5–0.57 h.d. Mandibles and mandibular processes as *E. brunettii.* No subcephalic setae, but six short cervical setae just posterior to cephalic capsule. Tail 2.5–3.0 a.b.d.

Spicules 320–345 μm (5.6–6.9 a.b.d.), transversely striated.

Gubernaculum in three parts: paired S-shaped parts 40–48 μm, dilated and knobbed distally, with an unpaired curved plate dorsal to these. Supplement large and *spatulate*, about half the spicule length in front of the cloaca. Testes paired, outstretched. Paired postcloacal papillae swollen in middle.

Vulva at 56–58% of body length.

Distribution. Exe estuary; Strangford Lough, Northern Ireland; Loch Etive, Scotland (intertidal fine to muddy sand).

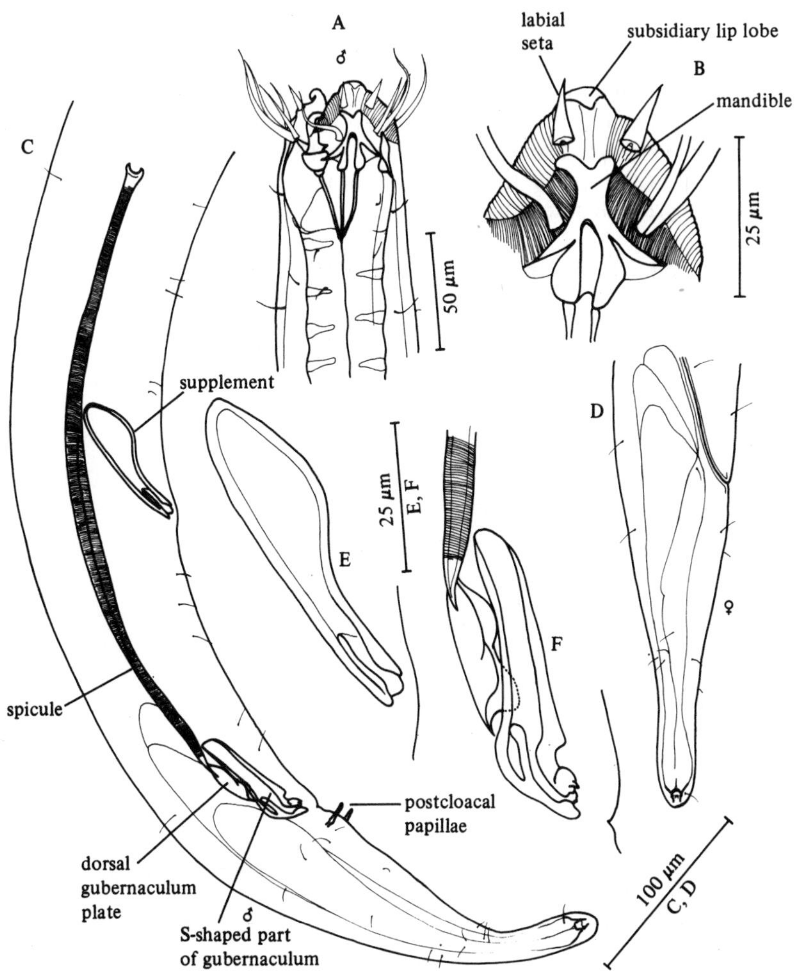

Fig. 47. *Enoploides longispiculosus*. A, Male head; B, Detail of lip region; C, Male tail; D, Female tail; E, Supplement; F, Gubernaculum and spicule tip. Original.

Enoploides spiculohamatus Schulz, 1932
(Fig. 48)

Description. Body length 2.7–2.8 mm. Maximum diameter 78–83 μm (a = 34). Very similar to *E. longispiculosus* except that the supplement is smaller and not so expanded at the proximal end, and is situated only one-quarter to one-third of the spicule length in front of the cloaca. Spicules 335–356 μm long, gubernaculum length 42–45 μm.

Distribution. Isle of Cumbrae, Scotland (medium coarse sand at 6 m depth).

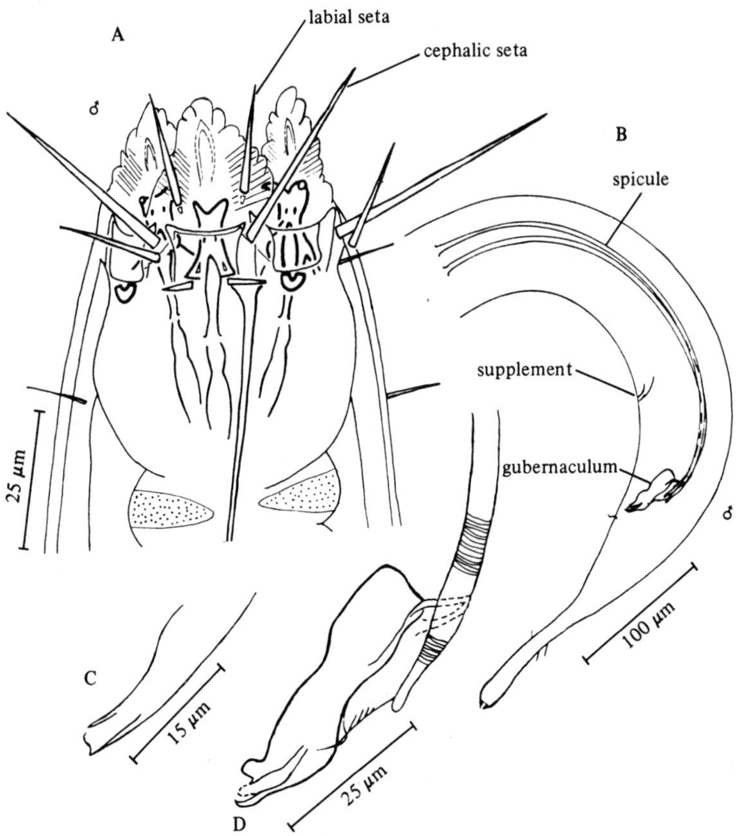

Fig. 48. *Enoploides spiculohamatus*. A, Male head; B, Male tail; C, Supplement; D, Gubernaculum and spicule tip (from Benwell, 1981).

Enoploides labrostriatus (Southern, 1914)
(Fig. 49)

Description. Body length 5.6–7.4 mm. Maximum diameter 132 μm (a = 41–56). Head structures similar to other species. Unlike the previous three species the tail has long terminal setae.

Spicules very long (490 μm) and said to be *smooth.*

Gubernaculum paired, with wide funnel-shaped proximal ends, distal parts curved with grooves bordered by two teeth. Supplement simple, tubular, 0.3 times spicule length in front of cloaca.

Distribution. Clew Bay, West Ireland (sublittoral sand and shells, 44 m depth).

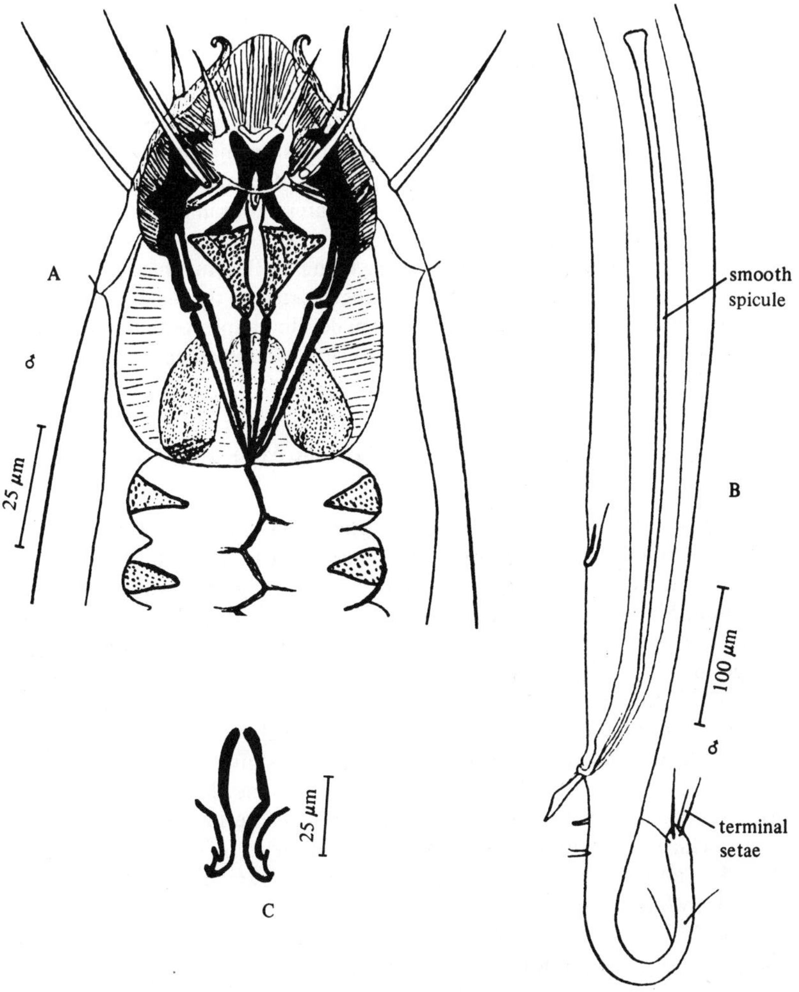

Fig. 49. *Enoploides labrostriatus*. A, Head; B, Male tail; C, Ventral view of gubernaculum (from Southern, 1914).

Genus ENOPLOLAIMUS De Man, 1893

Keys: Wieser, 1953; Hopper, 1962

Enoplolaimus species are characteristically found in sandy sediments in shallow water. They are voracious predators, specimens often being found ingesting other nematodes whole. Recurved points projecting into the buccal cavity from the tips of the mandibles prevent the prey from escaping. All species have mandibles which appear as two lateral rods united by an anterior curved bar, the cephalic setae arise from near the base of the cephalic capsule, and the spicules are never very elongate. Useful characters distinguishing the species are the relative lengths of setae on the head, the number and arrangement of cervical setae in the male, the shape of the tail, the form of the spicules and the position of the precloacal supplement.

Species : 28

Enoplolaimus propinquus De Man, 1922
(Fig. 50)

Description. Body length 1.6–2.6 mm. Maximum diameter 47–85 μm (a = 29–35). Cuticle with fine transverse striations. Head tapered, often bent to one side in fixed specimens. Three fairly low flap-like lips with no internal striation. Six labial setae 12–15 μm. Six long and four short cephalic setae, level with base of cephalic capsule: the single lateral setae 56–72 μm, *slightly* longer than the longer setae of each submedian pair; shorter submedian setae about half the length of the longer ones. Four small sublateral setae just posterior to the cephalic setae. Male with a circle of 8 long cervical setae just behind head, and a particularly dense area of shorter body setae about 3.5 h.d. from anterior. Female with no cervical setae. Mandibles typical of genus, appearing as two lateral rods united by an anterior curved bar. Teeth (or onchia) equal in size. Oesophagus cylindrical, 0.20–0.25 times body length. Tail 4–6 a.b.d. long, anterior half conical, remainder cylindrical, tip slightly swollen.

Spicules equal, 38–48 μm (curve), bent sharply in middle with a single laterally projecting tooth at their distal ends.

Gubernaculum paired, straight, 15–19 μm. Supplement 9–11 μm, a simple cuticularised cylinder positioned ventrally 1.9–2.3 a.b.d. in front of cloaca.

Ovaries symmetrical, paired, reflexed. Vulva at 59–60% of body length.

Distribution. Northern Ireland; Exe estuary; Loch Ewe (Scotland); Clyde (all intertidal sand).

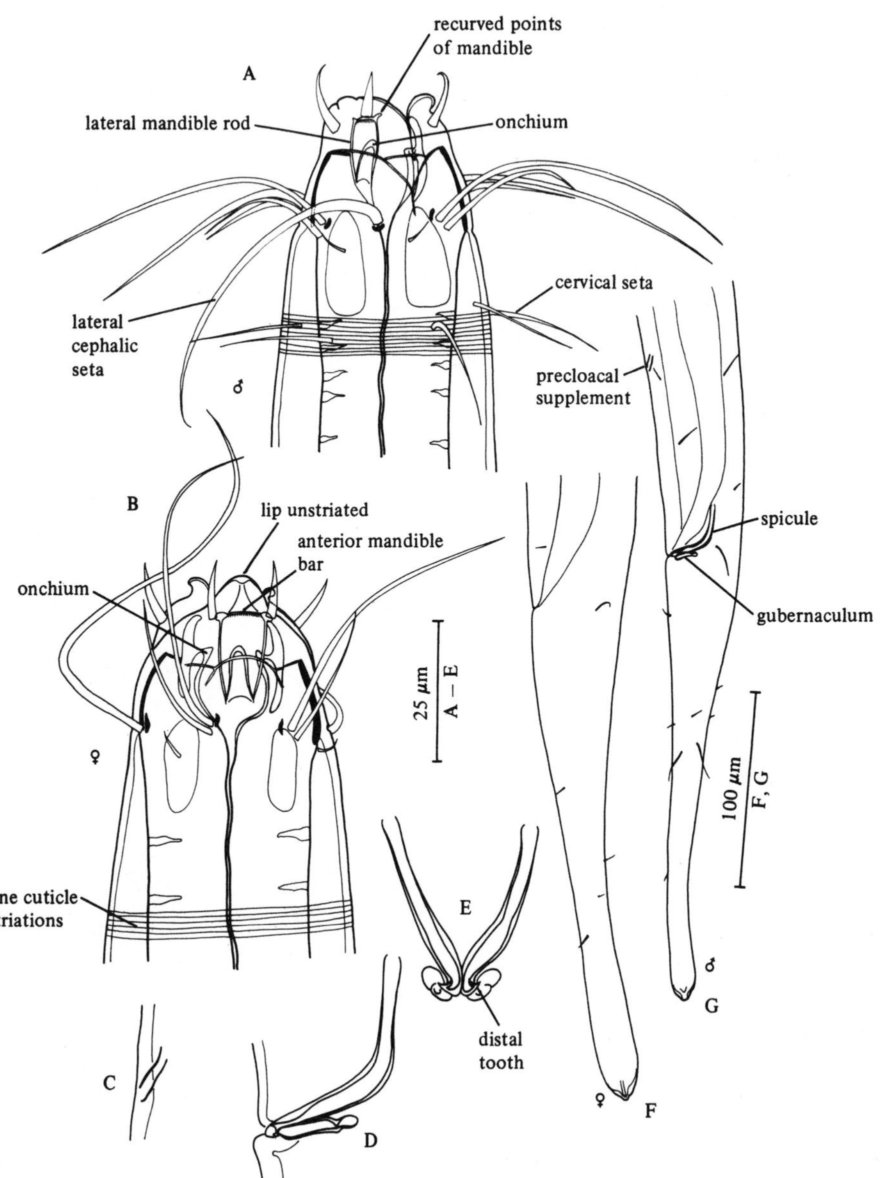

Fig. 50. *Enoplolaimus propinquus*. A, Male head; B, Female head; C, Supplement; D, Spicules and gubernaculum (lateral); E, Spicules and gubernaculum (ventral); F, Female tail; G, Male tail. Original.

Enoplolaimus denticulatus Warwick, 1970
(Fig. 51)

Description. Body length 3.4–4.6 mm. Maximum diameter 60–86 μm (a = 53–61). Cuticle smooth. Inner surfaces of lips with semi-lunar striations. Six long cephalic setae fairly stout, equal in length, up to 2 h.d. long: four shorter submedian setae more slender, up to 0.9 h.d. Numerous long cervical setae in male, up to 1 h.d. long. Female with single ring of six short cervical setae. Tail 4.0–5.5 a.b.d.

Spicules straight, 53–60 μm long, distal tip strongly hooked and bearing fine denticles.

Supplement 2.8–3.3 a.b.d. in front of cloaca.

Vulva at 60–67% of body length.

Distribution. Exe estuary; Isles of Scilly (intertidal sand).

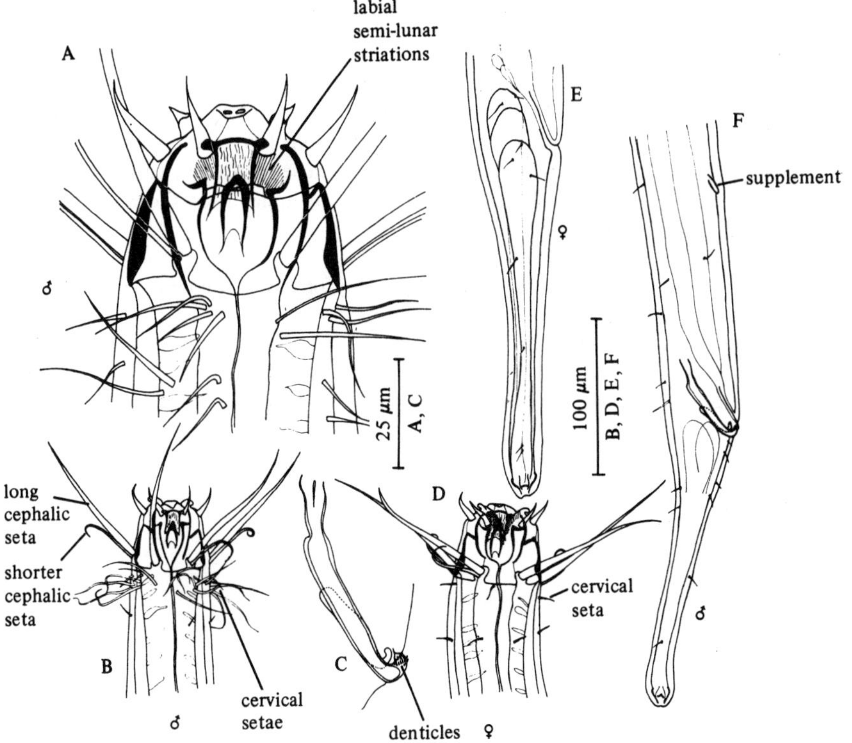

Fig. 51. *Enoplolaimus denticulatus*. A, Detail of male head; B, Male head; C, Spicules and gubernaculum; D, Female head; E, Female tail; F, Male tail. Original.

Enoplolaimus litoralis Schulz, 1936
(Fig. 52)

Description. Body length 1.9–2.2 mm. Maximum diameter 37–68 μm (a = 30–55). Cuticle smooth. No striation on lips. Two long lateral cephalic setae 2.9–3.0 h.d. in the male, 2.2–2.6 in the female; four longer submedian setae *about half this length*, the four shorter submedian setae about two-thirds the length of the longer. Ten long cervical setae in male, four short setae in female. Tail 4.1–5.0 a.b.d.

Spicules arcuate, 43–51 μm long (curve).

Gubernaculum with small dorsocaudal apophysis. Supplement 2.0–2.3 a.b.d. in front of cloaca.

Vulva at 56–63% of body length.

Distribution. Exe estuary; Isles of Scilly (intertidal sand).

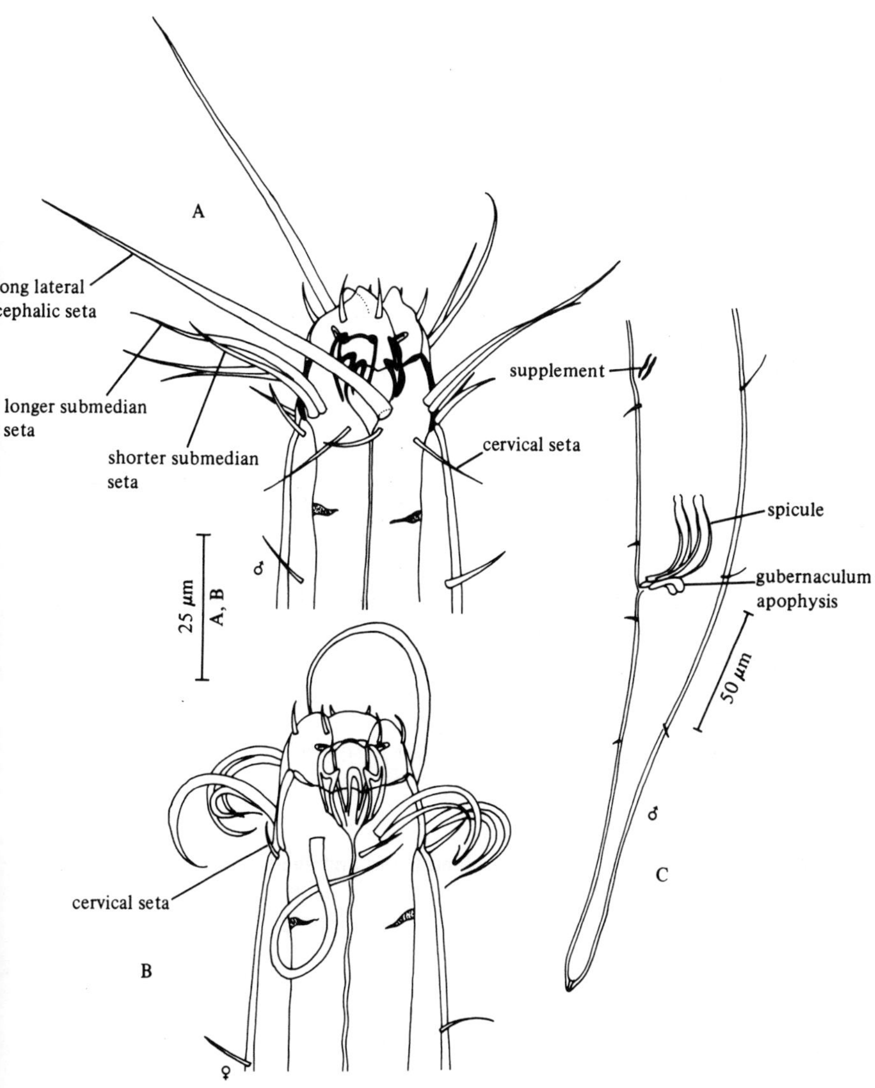

Fig. 52. *Enoplolaimus litoralis*. A, Male head; B, Female head; C, Male tail. Original.

Enoplolaimus vulgaris De Man, 1893
(Fig. 53A, B)

Description. Body length 3.1–5.6 mm (a = 55–75). Cuticle finely striated. Long lateral cephalic setae 1.6–1.8 h.d., longer submedian setae a little shorter, shorter submedian setae about two-thirds as long. Male with eight cervical setae arranged in pairs of unequal length, the longest about two-thirds of h.d. Female with four short cervical setae. Tail a little more than 4 a.b.d.

Spicules arcuate, 55 μm long.

Gubernaculum with dorsally directed apophysis. Supplement 2.5–3.5 a.b.d. in front of cloaca.

Vulva at 54% of body length.

Distribution. Falmouth (intertidal sand).

Enoplolaimus longicaudatus (Southern, 1914)
(Fig. 53C, D)

Description. Body length 2.9–3.6 mm. Maximum diameter 50–76 μm (a = 40–62). Cuticle with fine transverse striation. Six long cephalic setae equal, about 1.3 h.d., four shorter setae about three-quarters as long. Males with a ring of six cervical setae longer than the cephalic setae. Cervical setae short in female. *Tail very long* (about 9.5 a.b.d.).

Spicules arcuate, 51–54 μm.

Supplement very small (6 μm) and *close to cloaca* (level with proximal ends of spicules).

Vulva at 57–62% of body length.

Distribution. Clew Bay, West Ireland (sand and shells at 44 m depth).

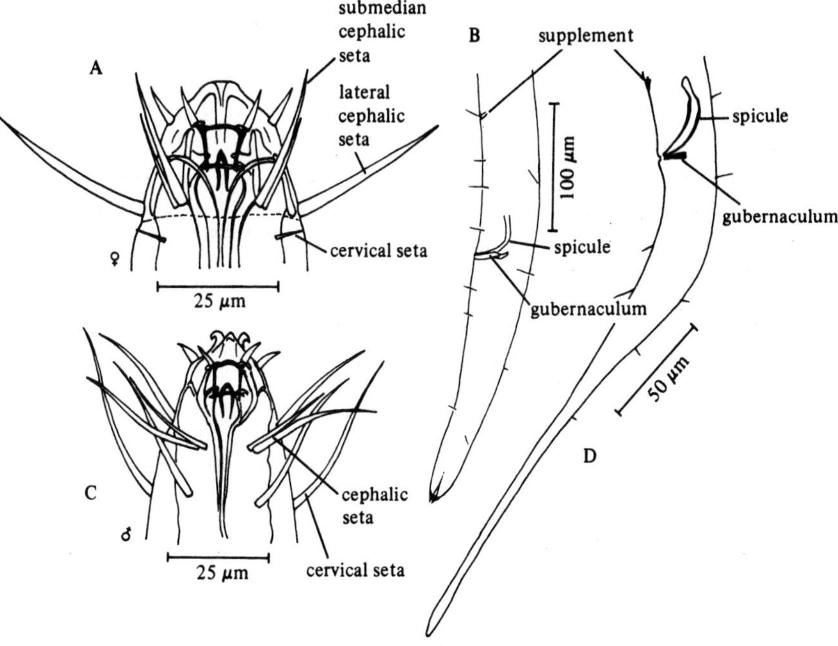

Fig. 53. *Enoplolaimus vulgaris*. A, Female head (from De Man, 1893); B, Male tail (from Riemann, 1966). *Enoplolaimus longicaudatus*. C, Male head; D, Male tail (from Southern, 1914).

Enoplolaimus subterraneus Gerlach, 1952
(Fig. 54)

Description. Body length 2.4–3.3 mm. Maximum breadth 62–110 μm (a = 30–39). Cuticle smooth. Lips with semi-lunar striations. Labial setae very short (7–8 μm) and stout. Six longer cephalic setae all 1.4 h.d. long, four shorter ones are only one-quarter as long. Male with six short cervical setae, none in female. Mandibles shorter and broader than in other members of the genus. Tail 2.7–3.8 a.b.d., strongly constricted behind the anus.

Spicules 38–45 μm, straight.

Supplement 2.1 a.b.d. in front of cloaca.

Vulva at 58–64% of body length.

Distribution. Exe estuary (intertidal sand).

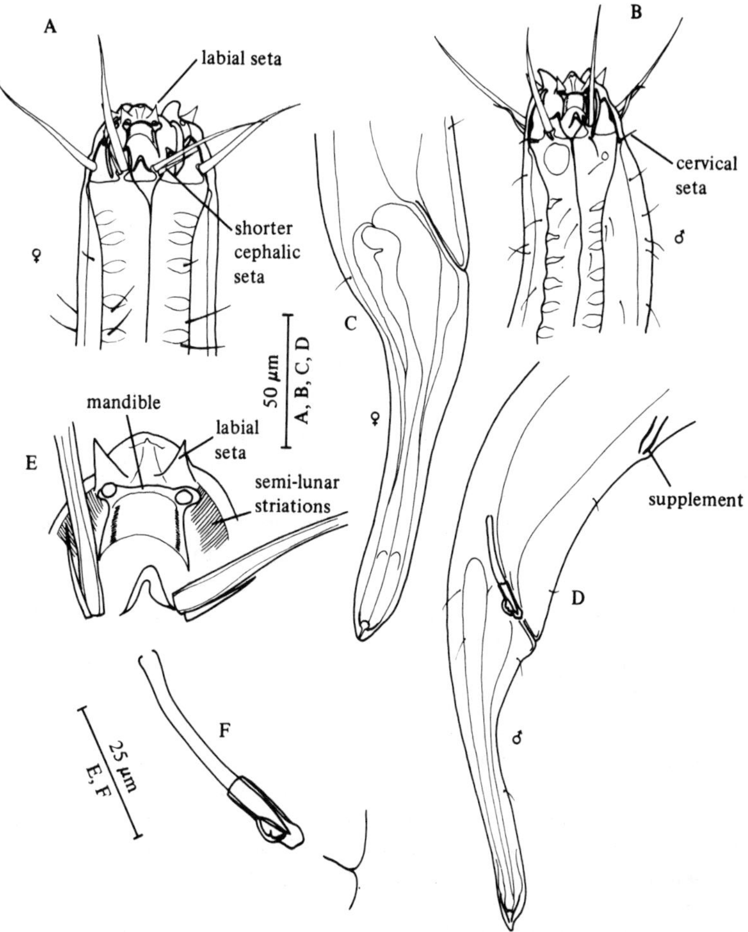

Fig. 54. *Enoplolaimus subterraneus*. A, Female head; B, Male head; C, Female tail; D, Male tail; E, Detail of lip region; F, Spicules and gubernaculum. Original.

Genus EPACANTHION Wieser, 1953

Key: Wieser, 1953
The species of *Epacanthion* so far recorded from the British Isles form a very heterogeneous group of species easily distinguished apart by the relative dimensions of the mandibles, the arrangement of setae on the head end and the structure of the male copulatory apparatus. The mandibles are intermediate in structure between those of *Enoploides* and *Enoplolaimus/Mesacanthion/Paramesacanthion* more solid in appearance than those of the latter group of species, but with a central clear sheet of cuticle unlike *Enoploides*.

Species: 19

Epacanthion gorgonocephalum Warwick, 1970
(Fig. 55)

Description. Body length 4.3–4.8 mm. Maximum diameter 82–138 μm (a = 35–52). Cuticle smooth. High lips with marked subsidiary lobe, internally marked with semi-lunar striations: outer parts with wider spaced striations, lip margin scalloped; subsidiary lobe with no internal striation or scalloped edge. Labial setae 19–23 μm. Six long (63–82 μm) and four short (32–43 μm) cephalic setae situated at level of bases of onchia. Male with twelve groups of subcephalic setae, three per group, one long, one medium and one short. Female with only six short subcephalic setae, or these may be entirely lacking. Mandibles appear as two longitudinal rods joined by a thin sheet of cuticle, relatively long and slender. Onchia equal in size. Onchial plates rounded posteriorly, each curving inwards to provide a cup-shaped base to the buccal cavity. Numerous long dense cervical setae in the male with a region of very dense but rather shorter setae about one-third of the way down the oesophagus length. Cervical setae sparse in female. Oesophagus 0.22 times body length, swollen anteriorly round base of buccal cavity. Tail 3.6–4.4 a.b.d. long in male, 3.1–3.4 in female, anterior half conical, posterior half cylindrical.

Spicules short, 50–53 μm (0.9 a.b.d.), slightly curved, with a pair of laterally curving knobs distally.

Gubernaculum a short double tube with two small lateral projections. No precloacal supplement.

Ovaries paired, symmetrical, opposed, reflexed. Vulva at 60% of body length.

Distribution. Exe estuary; Isles of Scilly (intertidal sand).

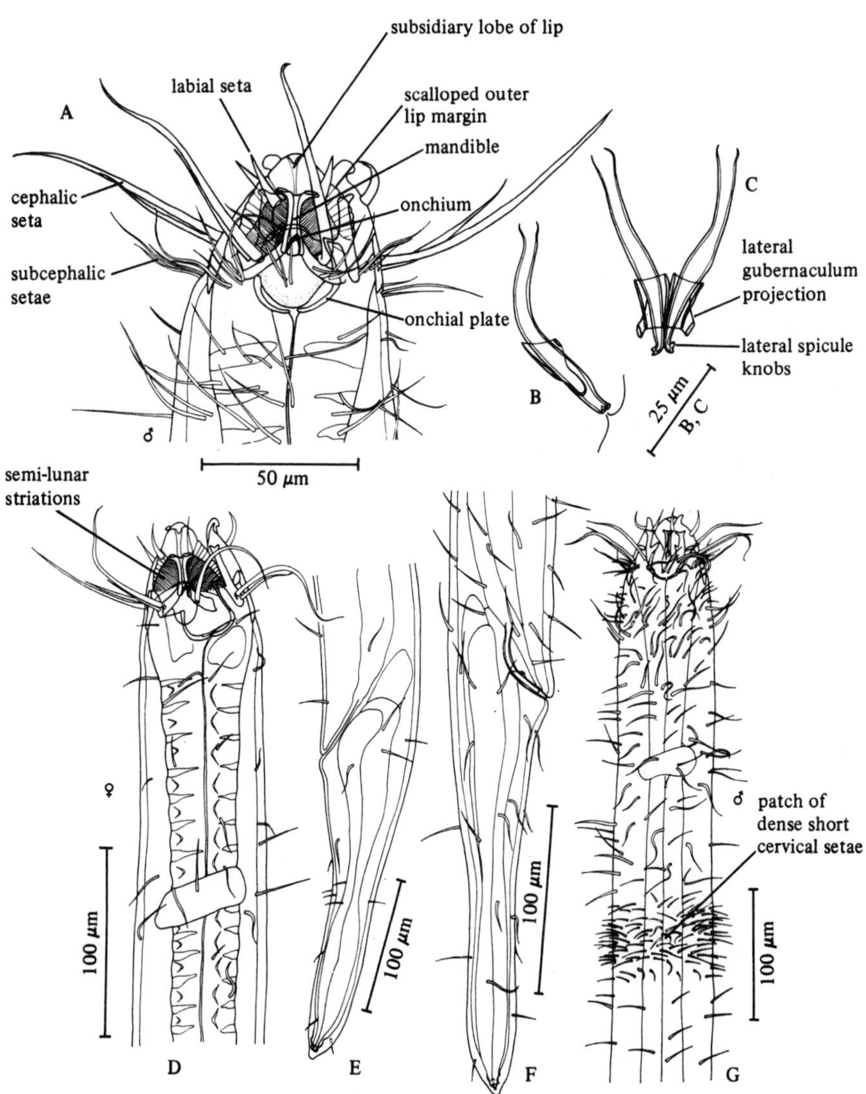

Fig. 55. *Epacanthion gorgonocephalum*. A, Male head; B, Spicules and gubernaculum (lateral); C, Spicules and gubernaculum (ventral); D, Anterior end of female; E, Female tail; F, Male tail; G, Anterior end of male. Original.

Epacanthion buetschlii (Southern, 1914)
(Fig. 56)

Description. Body length 4.0–4.4 mm. Maximum diameter 130–180 μm (a = 23–34). Cuticle with very fine transverse striation. Longest cephalic setae 0.6 h.d. A circle of long slender subcephalic setae near the base of the cephalic capsule. Numerous long slender cervical setae, longer and more numerous in males than females. Mandibles of two lateral strongly cuticularised rods with a central lacuna, shorter and broader than in *E. gorgonocephalum.* Tail 4 a.b.d.

Males with a pair of short stout submedian spines just behind cloaca. Spicules elongate, 480 μm (9.6 a.b.d.), transversely striated.

Gubernaculum paired, simple rods. Precloacal supplement 32 μm long, one tail length in front of cloaca.

Distribution. Clew Bay and Dingle Bay, West Ireland (sublittoral sand).

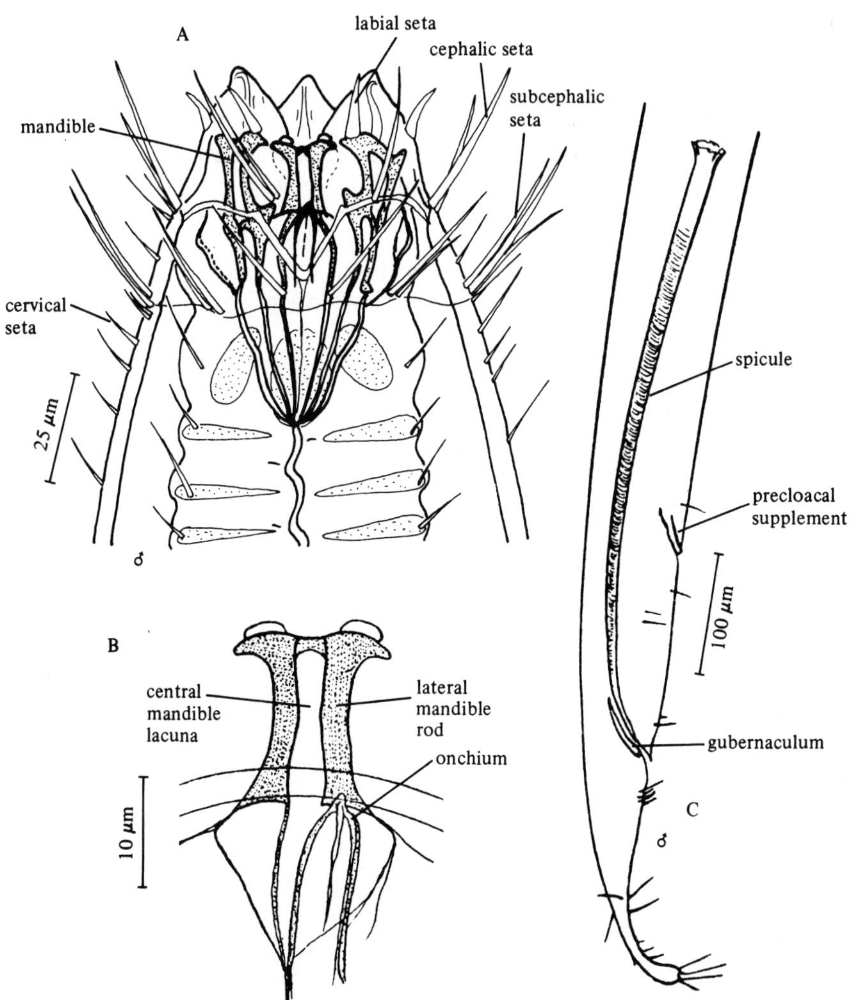

Fig. 56. *Epacanthion buetschlii*. A, Male head; B, Detail of mandible and onchium; C, Male tail (from Southern, 1914).

Epacanthion mawsoni Warwick, 1977
(Fig. 57)

Description. Body length 5.1–7.0 mm. Maximum diameter 52–97 μm (a = 72–97). Cuticle smooth. Longest cephalic setae 0.70–0.87 h.d. Twelve pairs of cervical setae just behind base of cephalic capsule in male, only four short setae in female. Mandibles of similar structure to preceding species but very short and broad. Dorsal onchium much smaller than the two subventral ones. Tail conical, 4.8–5.4 a.b.d.

Spicules short, 59 μm (1.2 a.b.d.), arcuate.

Gubernaculum a small triangular plate. Supplement 7.5 μm long, 0.4 times tail length in front of cloaca.

Distribution. Isles of Scilly (coarse sublittoral sediment).

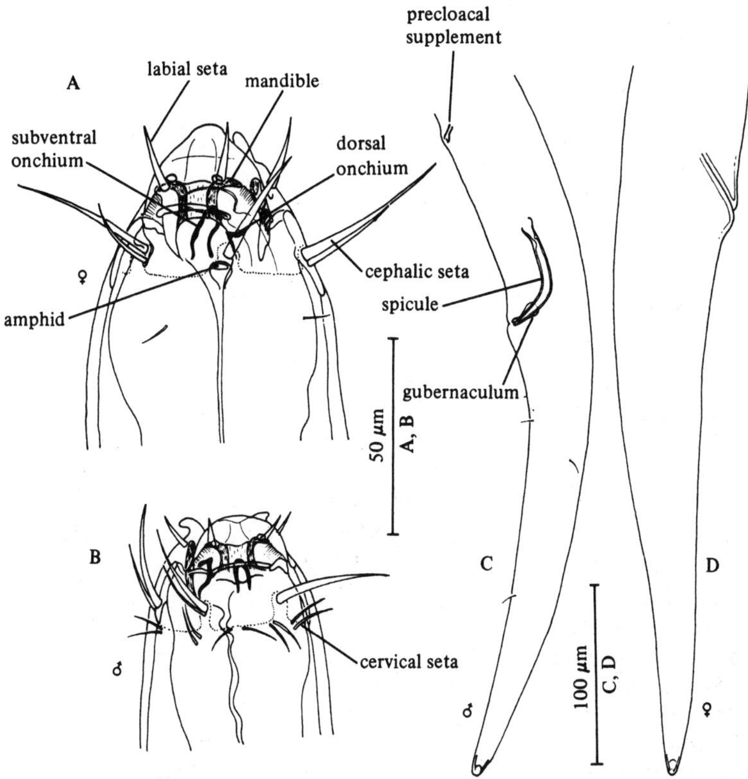

Fig. 57. *Epacanthion mawsoni*. A, Female head; B, Male head; C, Male tail; D, Female tail. Original.

Genus MESACANTHION Filipjev, 1927

Key: Wieser, 1953
Mesacanthion has a similar mandible structure to *Enoplolaimus*, but the cephalic setae arise from the middle or anterior end of the cephalic capsule. Unlike *Enoplolaimus*, several species have elongated spicules. The three species so far recorded from the British Isles are all distinctive, the main distinguishing characters being the same as for *Enoplolaimus*.

Species : 37

Mesacanthion africanthiforme Warwick, 1970
(Fig. 58)

Description. Body length 2.6–4.5 mm. Maximum diameter 33–66 μm (a = 56–82). Cuticle smooth. Some irregular areas of pigmentation just posterior to cephalic capsule. Head characteristically dome-shaped. Three lips relatively low with no internal striations. Six labial setae stout and conical, 0.2 h.d. Cephalic setae arise from about the middle of the cephalic capsule, the posterior border of the capsule being deeply incised and the bases of the cephalic setae accommodated in fenestrae. Longer six cephalic setae 0.6–0.9 h.d., shorter four about half their length. Six files of longish setae extend one-third to two-thirds of the way down the oesophagus length in both sexes, more numerous in male. Mandibles as *Enoplolaimus*, with two lateral rods united by an anterior curved bar. Teeth (onchia) large and equal in size. Oesophagus cylindrical, 0.18–0.24 times body length. Tail 3.9–4.9 a.b.d. long, conical for anterior third, remainder cylindrical.

Spicules small (0.6–0.75 a.b.d., 20–33 μm), fairly straight, pointed and open ended distally with a thickened cuticular ring just proximal to the tip.

Gubernaculum 10–13 μm, a pair of tubes surrounding the distal ends of the spicules, tubes united by a median bar appearing roughly triangular in lateral view. No precloacal supplement. Male tail with two pairs of stout conical spines posterior to cloaca which probably have a copulatory function.

Ovaries paired, symmetrical and doubly reflexed. Vulva at 59–62% of body length.

Distribution. Exe estuary; Isles of Scilly (intertidal sand).

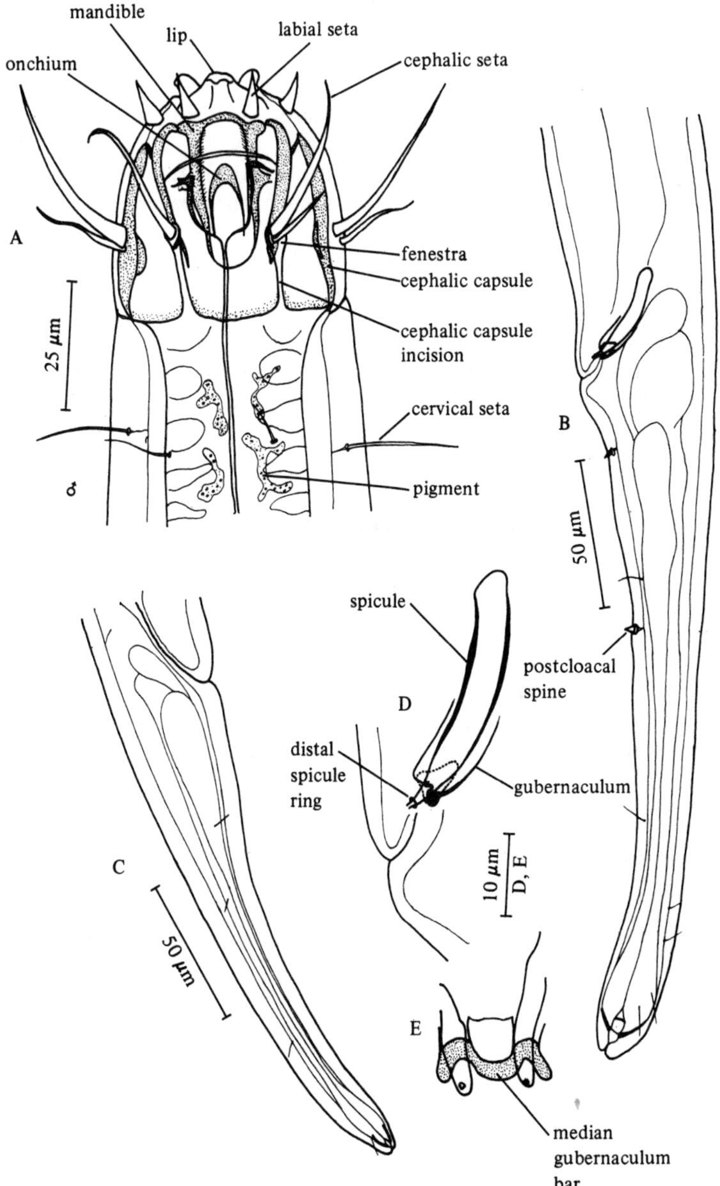

Fig. 58. *Mesacanthion africanthiforme*. A, Male head; B, Male tail; C, Female tail; D, Spicules and gubernaculum (lateral); E, Spicules and gubernaculum (ventral). Original.

Mesacanthion diplechma (Southern, 1914)
(Fig. 59)

Description. Body length 3.3–3.7 mm. Maximum diameter 76–128 μm (a = 27–44). Cuticle with fine transverse striation. Cephalic setae arise from anterior border of cephalic capsule, longer six 1.1 h.d., shorter four two-thirds this length. Males with six pairs of cervical setae just posterior to the cephalic capsule, and six files of four setae further posterior. These setae absent in female. Tail 4.4–4.8 a.b.d.

Spicules very unequal in length and shape. Right spicule elongate, 500–510 μm (8.3–9.4 a.b.d.), transversely striated. Left spicule short (1.25 a.b.d.) and smooth, in two halves with a suture between them; proximal half with strong dorsal hook at its distal end.

Gubernaculum in two halves, each with a long posterior apophysis. Supplement 1.3–1.5 a.b.d. in front of cloaca. Male tail with ventral setose papilla near the tip.

Vulva just behind middle of body.

Distribution. Clew Bay, West Ireland (sublittoral sand); Northumberland (sublittoral mud).

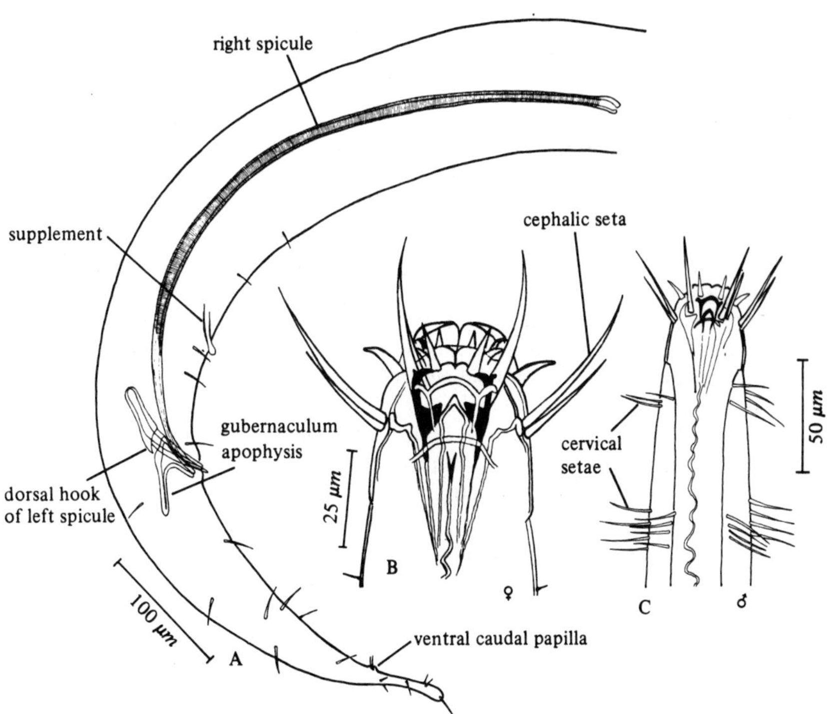

Fig. 59. *Mesacanthion diplechma*. A, Male tail (original); B, Female head; C, Male head (from Southern, 1914).

Mesacanthion hirsutum Gerlach, 1952
(Fig. 60)

Description. Body length 1.9–2.3 mm. Maximum diameter 35–53 μm (a = 37–63). Cuticle smooth. Lips very high and pointed. Six labial setae unusually long and slender (12–14 μm, about half the h.d.). Cephalic setae arise slightly anterior to the middle of the cephalic capsule, longer six 1 h.d., shorter four about one-third this length. Six whorls of long slender cervical setae in male, the first just posterior to the cephalic capsule, posterior two close together at about the level of the nerve ring. First whorl absent in female, setae of remaining whorls less numerous and shorter than in the male, posterior two whorls with only three or four setae. Dorsal onchium is slightly smaller than the two subventral ones. Tail relatively long, about 8 a.b.d.

Spicules 1.3 a.b.d. (36–42 μm), arcuate, slender.

Supplement 2.6–3.3 a.b.d. in front of cloaca.

Vulva at 51–54% of body length.

Distribution. Exe estuary (intertidal sand).

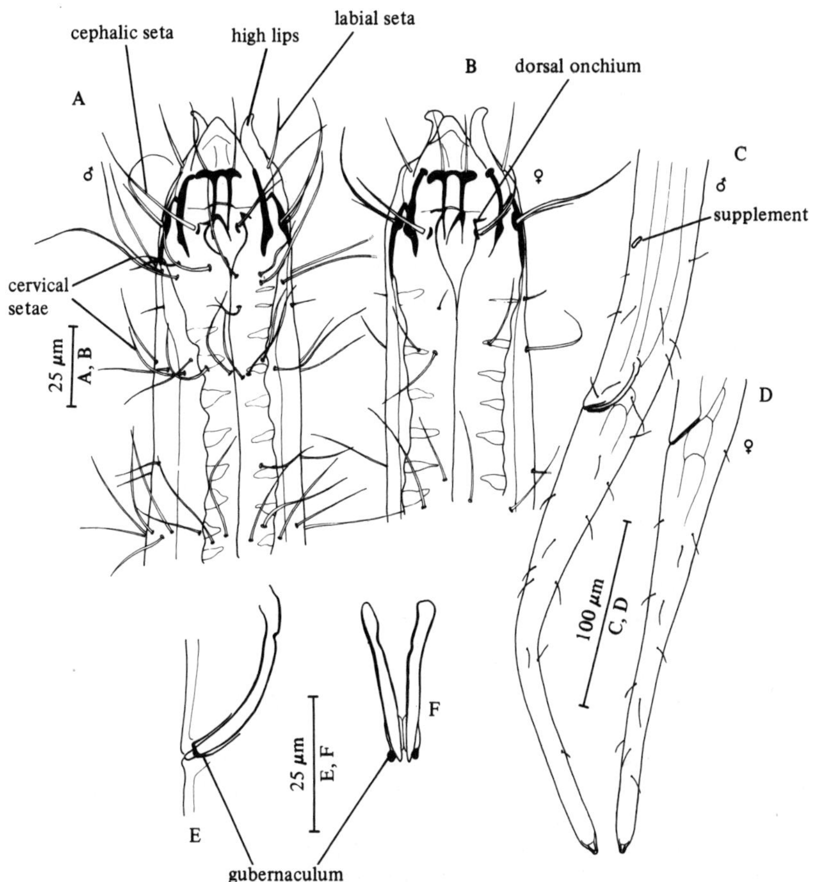

Fig. 60. *Mesacanthion hirsutum*. A, Male head; B, Female head; C, Male tail; D, Female tail; E, Spicules and gubernaculum (lateral); F, Spicules and gubernaculum (ventral). Original.

Genus PARAMESACANTHION Wieser, 1953

Keys: Wieser, 1953; Warwick, 1970

Paramesacanthion appears to be confined to offshore sediments and species are seldom found in shallow water or on the shore. The genus is defined by the extreme anterior position of the cephalic setae and the double-jointed spicules, although this latter feature is sometimes difficult to discern in the short-spiculed species. In all species there is strong sexual dimorphism in the setation of the head end, and females are almost impossible to distinguish apart with certainty. Characters of most value in identifying males are the head setation and the form of the reproductive apparatus.

Species : 11

Paramesacanthion hirsutum Warwick, 1970
(Fig. 61)

Description. Body length 2.6–2.8 mm. Maximum diameter 76–93 μm (a = 27–35). Cuticle with fine transverse striation. Three lips relatively low with faint semi-lunar striations. Labial setae small, conical. Cephalic setae situated at level of tips of mandibles, well anterior to the cephalic capsule. In male, six longer cephalic setae 0.4 h.d., shorter four about half this length. Longer cephalic setae in female only 0.2 h.d., shorter three-fifths as long. Male with four subcephalic setae just posterior to the submedian pairs of cephalic setae, and ten much longer subcephalic setae at level of the tips of the onchia: eight dense groups of cervical setae, 2–12 setae per group, just posterior to cephalic capsule. In females and juveniles there are no subcephalic or cervical setae. Prominent oval 'cephalic organs' present near anterior end of mandibles, sublateral in position. Mandibles the same form as *Enoplolaimus* and *Mesacanthion.* Onchia large, equal. Cephalic capsule with deep incisions extending from posterior to anterior border, anterior part of each incision bordered by slender cuticular ribs, with similar ribs along anterior border of capsule forming the cephalic ring. Oesophagus cylindrical, 0.23 times body length. Tail 4.6–4.8 a.b.d. long.

Spicules equal, 48–53 μm (1.1–1.2 a.b.d.), arcuate, divided into two halves by a fine suture. No precloacal supplement.

Distribution. Northumberland (sublittoral fine sand and mud).

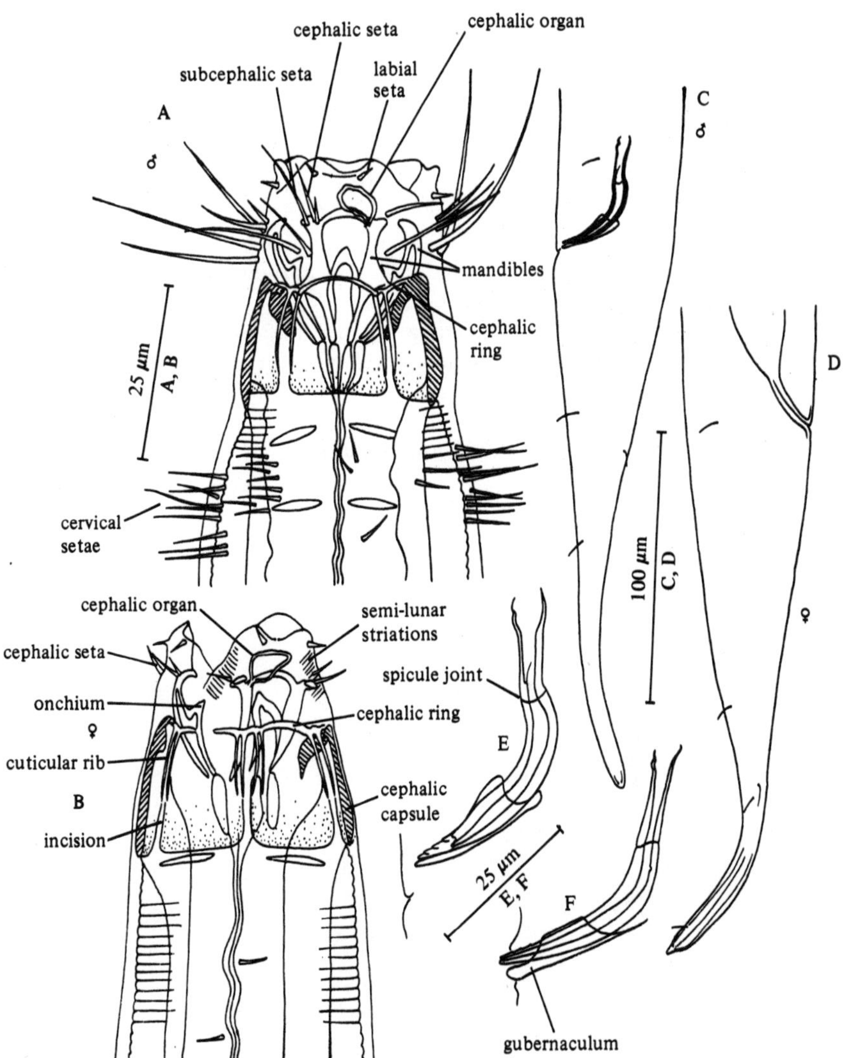

Fig. 61. *Paramesacanthion hirsutum*. A, Male head; B, Female head; C, Male tail; D, Female tail; E, F, Spicules and gubernaculum of two males. Original.

Paramesacanthion marei Warwick, 1970
(Fig. 62)

Description. Body length 2.7–2.9 mm. Maximum diameter 86–101 μm (a = 27–34). Arrangement of labial, cephalic and subcephalic setae as in *M. hirsutum*. However, in the male the eight groups of cervical setae are situated in a much more posterior position (0.17 times the oesophagus length from the anterior): there are six per group, 6 μm long. Subcephalic and cervical setae absent in female. Tail about 3 a.b.d. long.

Spicules equal, elongate, 2.2 a.b.d. (102–104 μm) divided into two joints with a sharp construction in the middle.

Gubernaculum elongate, extending beyond the spicular constriction. Large tubular precloacal supplement level with, or slightly anterior to, spicule tips.

Distribution. Northumberland (sublittoral mud).

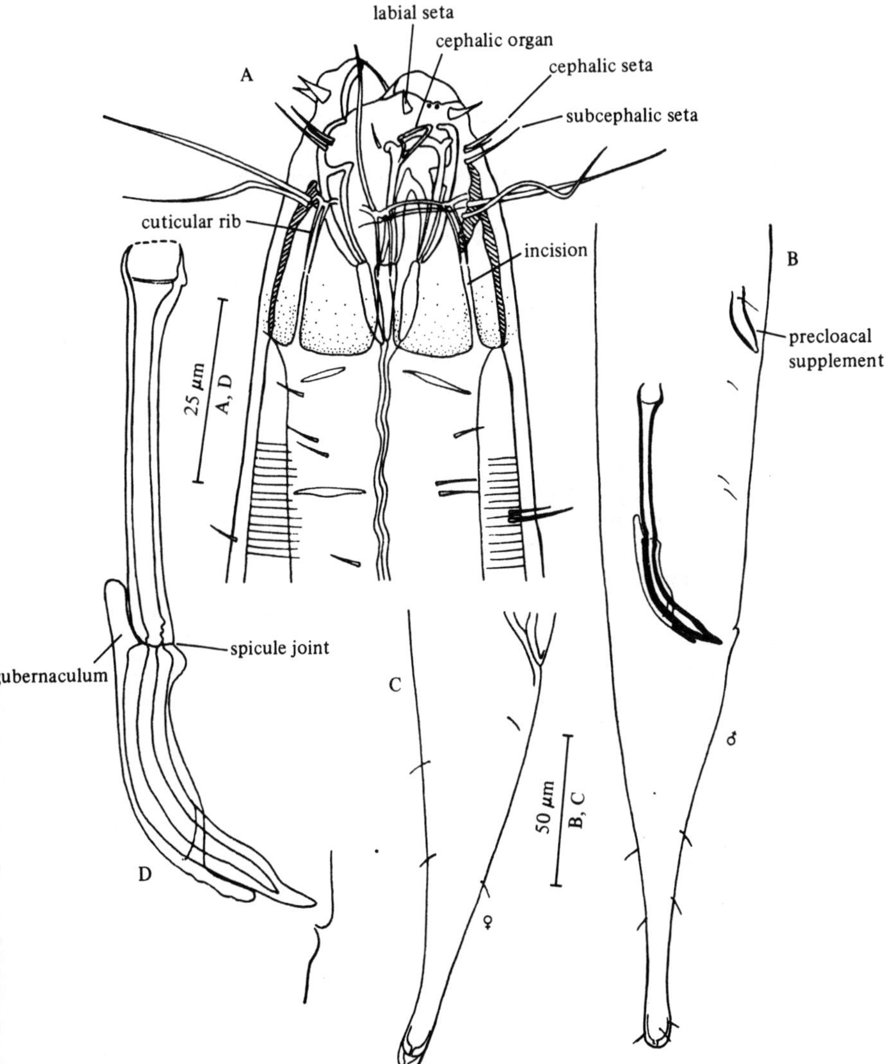

Fig. 62. *Paramesacanthion marei*. A, Male head; B, Male tail; C, Female tail; D, Spicules and gubernaculum. Original.

Genus THORACOSTOMOPSIS Ditlevsen, 1918

This genus is very distinctive in that the three onchia are very elongated and wrapped around one another to form an axial spear-like structure, whilst the mandibles are rather reduced. The posterior border of the cephalic capsule is strongly lobed and the incisions are very broad. Species are distinguished on the form of the spicules (long and slender or short and stout), the presence or absence of a gubernaculum, the shape of the posterior lobes of the cephalic capsule, and the length of the cephalic setae.

Species : 6

Thoracostomopsis doveae Warwick, 1970
(Fig. 63)

Description. Body length 5.4–5.5 mm. Maximum diameter 84–90 μm (a = 61–64). Cuticle with fine, faint transverse striation. Lips not strongly separated as in other members of Thoracostomopsidae. Labial setae 6–7 μm. Six longer cephalic setae 38–40 μm (2.1 h.d.); shorter four 8 μm (males), but relatively longer in juveniles. Male with a series of long slender cervical setae just posterior to cephalic capsule. Onchial spear-like structure jointed, anterior section 58 μm, posterior 34 μm, accommodated in anterior end of oesophagus. Three mandibles surround onchia at anterior end, each with a single median apophysis in addition to the lateral rods. Six posterior lobes of cephalic capsule irregular in shape, with pronounced punctations. Oesophagus more or less cylindrical, 0.13 times body length. Tail 3.1–3.4 a.b.d.

Spicules 100–103 μm, slender, cephalic proximally.

No gubernaculum. Tubular precloacal supplement level with spicule ends.

Distribution. North East England (*Laminaria* holdfasts).

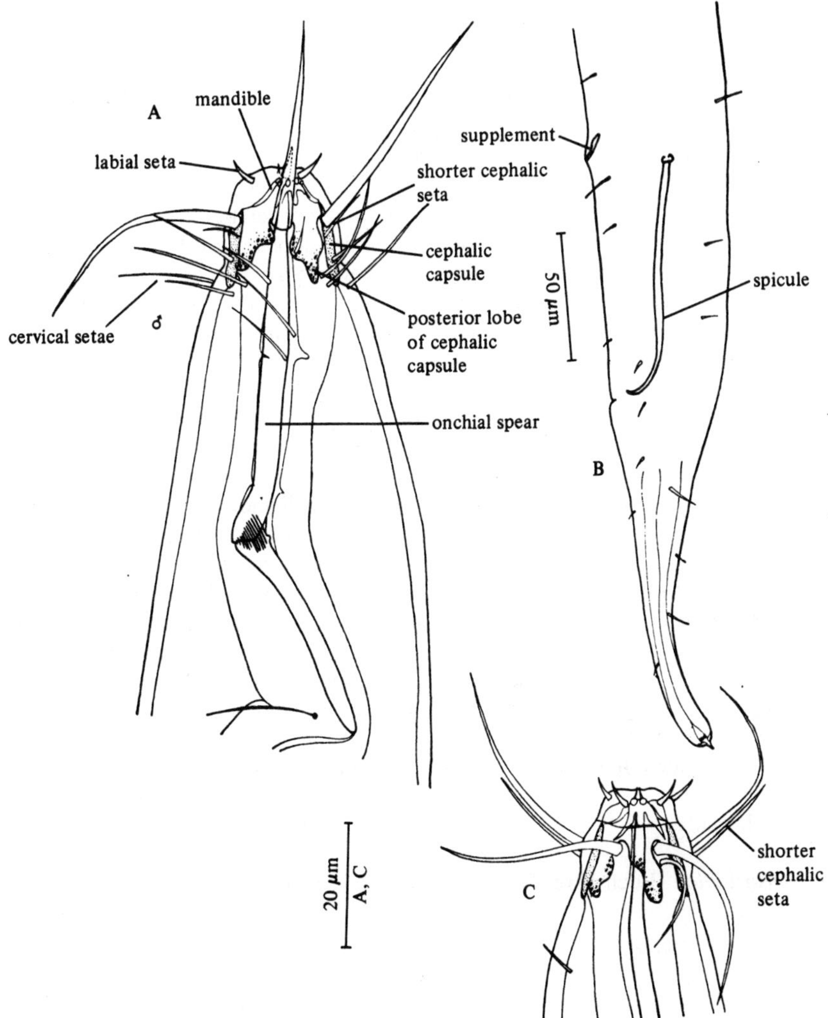

Fig. 63. *Thoracostomopsis doveae*. A, Male head; B, Male tail; C, Juvenile head. Original.

Genus TRILEPTIUM Cobb, 1933

Key: Wieser, 1959

This genus is similar to *Enoplolaimus* in the arrangement of cephalic setae relative to the cephalic capsule, but the lips are much lower and the mandibles and onchia are smaller and occupy a more anterior position. The development of mandibles and onchia is variable; in some species neither are visible and in others the onchia are strongly developed but the mandibles are weak. In some species there is differential development of the three onchia, with the dorsal one either larger or smaller than the other two. Other characters used to separate species are the form of the spicules and gubernaculum and the presence or absence of a precloacal supplement.

Species : 9

Trileptium parisetum Warwick and Platt, 1973
(Fig. 64)

Description. Body length 3.8 mm. Maximum diameter 41 μm (a = 100). Cuticle with faint transverse striation. Labial setae 7 μm. Cephalic setae 53 and 15 μm. Male with six pairs of subcephalic setae just posterior to the cephalic capsule, 35 and 17 μm long. Female not known. Ten pairs of cervical setae more posteriorly, 36 and 12 μm long. Mandibles and onchia set well forward in buccal cavity, equal in size. Mandibles small with a solid appearance, onchia relatively large. Oesophagus cylindrical, 0.19 times body length. Tail 4 a.b.d., anterior half conical, posterior half cylindrical.

Spicules 35 μm (0.85 a.b.d.), slender.

Gubernaculum 14 μm in two halves, each terminating distally in a small plate bearing two lateral projections. Precloacal supplement 49 μm in front of cloaca.

Distribution. Loch Ewe, Scotland (intertidal sand).

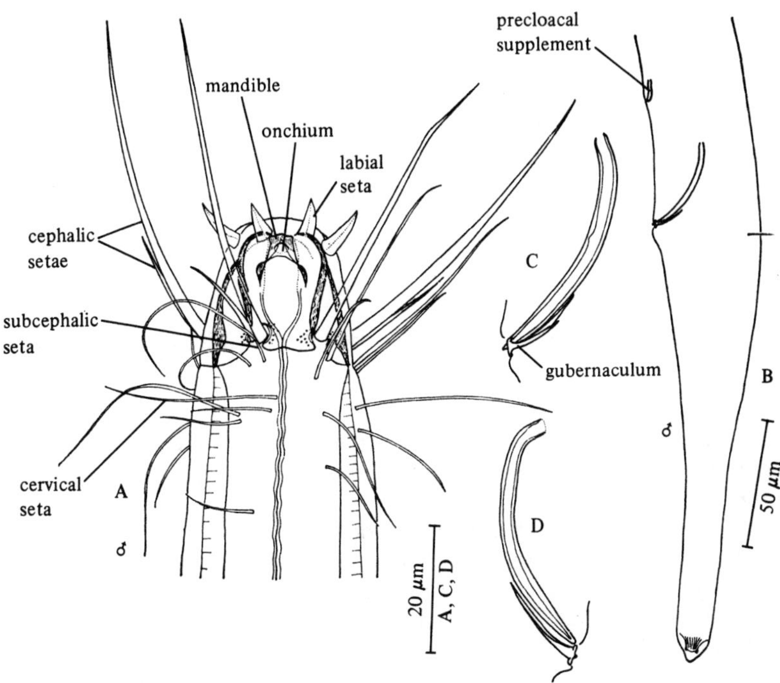

Fig. 64. *Trileptium parisetum*. A, Male head; B, Male tail; C, D, Spicules and gubernaculum of two males. Original.

Genus ANOPLOSTOMA Bütschli, 1874

Key: Wieser, 1953
Anoplostoma species are characterised by the large cylindrical buccal cavity which is not surrounded by oesophageal tissue and the presence of a copulatory bursa in the male. *Anoplostoma viviparum* is the most commonly reported species of the genus and the only one known from British waters. The species are distinguished primarily on the length of the cephalic setae, position of the amphids and length of the spicules.

Species : 15

Anoplostoma viviparum (Bastian, 1865)

(Fig. 65)

Description. Body length 1.3–2.1 mm. Maximum diameter 40–80 μm (a = 22–34). Body broadest in middle and attenuated at anterior and posterior ends. Cuticle smooth. Somatic setae absent. Six labial papillae. Six 8–11 μm (0.8–1.0 h.d.) cephalic setae situated just anterior to four 3–6 μm shorter setae. Amphids about 3 h.d. from anterior and 0.2 times c.d. wide. Buccal cavity cylindrical; no teeth. Oesophageal tissue does not surround the buccal cavity. Oesophagus increases in width posteriorly but no bulb. Nerve ring at 50–60% of oesophagus length. Tail long and tapering, 5–10 a.b.d.

Spicules 53–63 μm (2–3 a.b.d.), elongate.

Gubernaculum short paired structures. Two longitudinal cuticular ridges (bursa copulatrix) supported by spines lie either side of the cloaca. Testes paired, opposed.

Vulva at about 50% of body length. Two opposed, reflexed ovaries. Eggs of mature females characteristically contain developing larvae.

Distribution. Falmouth (estuarine mud); Blyth estuary (mud flat); Essex coast (intertidal mud); Exe estuary (intertidal mud); West Scotland (shallow subtidal sand); Strangford Lough, Northern Ireland (intertidal sand); Southampton Water (estuarine mud).

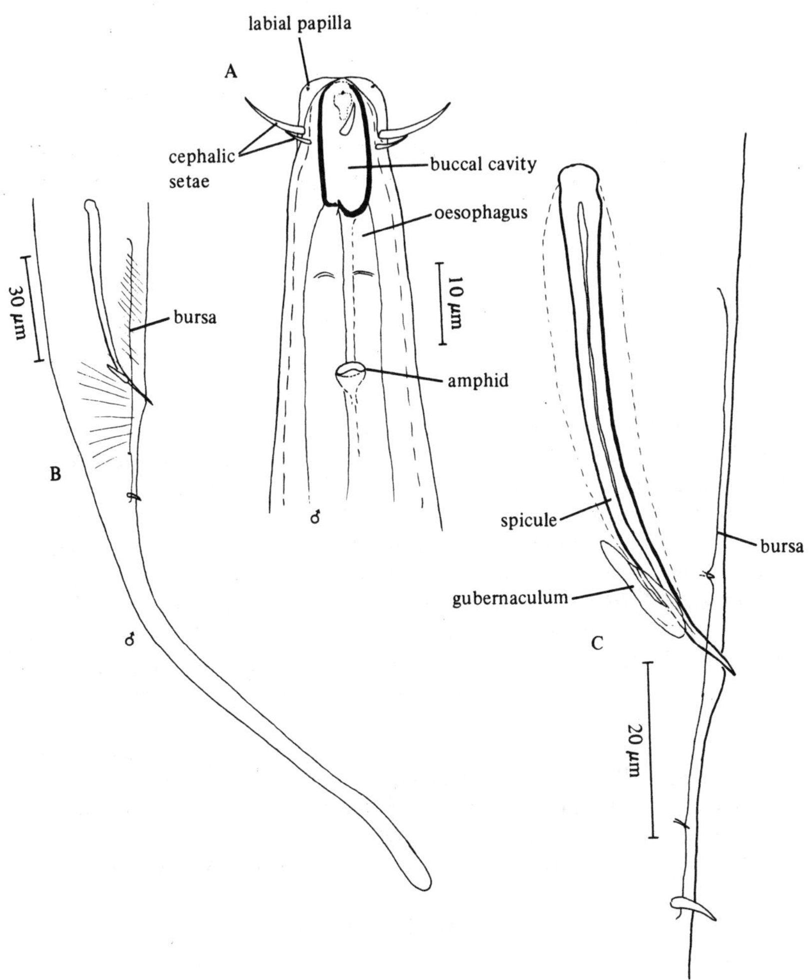

Fig. 65. *Anoplostoma viviparum*. A, Male head; B, Male tail; C, Copulatory apparatus. Original.

Genus CHAETONEMA Filipjev, 1927

Key: Platt, 1973

The genus is similar to *Anoplostoma* in having a large unarmed buccal cavity which is surrounded by oesophageal tissue only in the posterior part. However, the extreme sexual dimorphism of the amphid is unique and the buccal cavity is flask-shaped, not tubular. Species of the genus may be distinguished on relative lengths of the cephalic setae and presence or absence of a precloacal supplement.

Species : 5

Chaetonema riemanni Platt, 1973
(Fig. 66)

Description. Body length 1.2–1.3 mm. Maximum diameter 25–33 μm (a = 40–50). Cuticle smooth for the most part, but striations can be detected in the oesophageal region. Three lips, each with two triangular papillae. Ten cephalic setae; six 30–33 μm (3 h.d.) situated just posterior to four 10–11 μm (1 h.d.) setae. Somatic setae absent. Amphid structure shows sexual dimorphism, being an elongated groove in the male but only a small pocket-shaped structure in the female. The male amphid starts about 30 μm from the anterior, is about 100 μm long and ends in a distinct cuticularised structure. The female amphid is situated about 50 μm from the anterior, is about 7 μm long and has a small anterior fovea. The buccal cavity is flask-shaped, lacks teeth or mandibles, is supported by three semi-circular cuticularised bars and is only surrounded by oesophageal tissue in the posterior part. The tail gradually tapers but has a terminal swelling.

Spicules 38–40 μm long and characteristically bent in the proximal quarter.

Gubernaculum is a simple 16 μm long rod.

Ovaries paired, opposed and reflexed. Vulva at 44% of body length.

Distribution. Strangford Lough, Northern Ireland (intertidal sand).

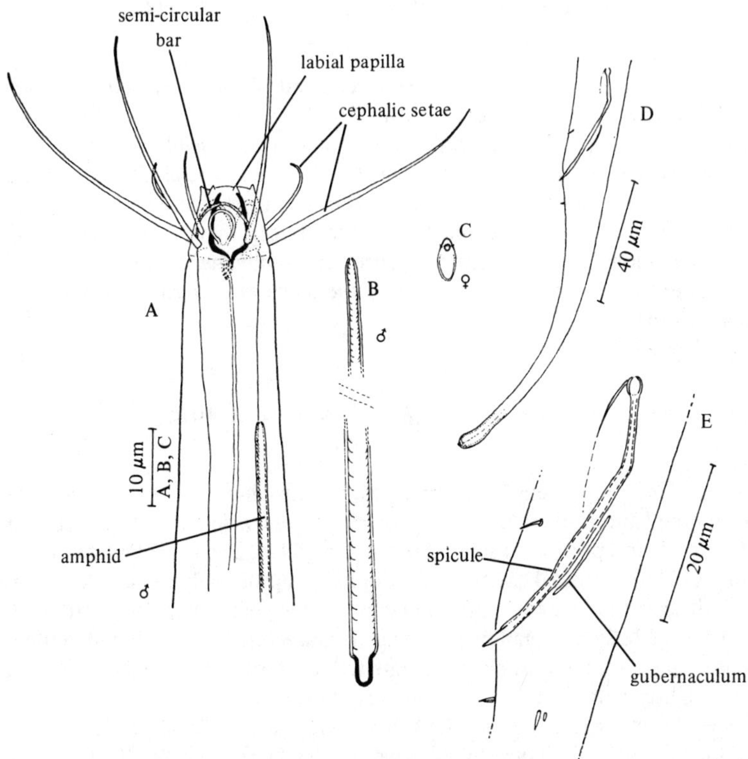

Fig. 66. *Chaetonema riemanni*. A, Male head; B, Male amphid; C, Female amphid; D, Male tail; E, Copulatory apparatus. Original.

Genus CRENOPHARYNX Filipjev, 1934

Key: Platonova, 1976

Crenopharynx species are only rarely reported, *C. marioni* being the only species found so far in the British Isles. Previous descriptions of this species are rather poor and none describe the amphid. It is possible that the animal characteristically comes to rest showing the dorsal or ventral side of the head, as did the specimen studied here. The genus is characterised by the combination of a weak cephalic capsule, buccal cavity with sclerotised structures, long spicules and no precloacal supplement. Species are differentiated mainly on the structure of the spicules and gubernaculum and shape of the tail.

Species : 9

Crenopharynx marioni (Southern, 1914)
(Fig. 67)

Description. Body length 4.3 mm. Maximum diameter 85 μm (a = 51). Attenuated anteriorly. Cuticle smooth and devoid of somatic setae except for tail and ventral precloacal setae. Six labial papillae. Ten cephalic setae 11 μm long (0.7 h.d.). Buccal cavity small. Cephalic capsule only weakly developed and difficult to distinguish. Oesophagus widens posteriorly and posterior to nerve ring becomes glandular giving a characteristic crenellated outline. Nerve ring 0.4 times oesophagus length from anterior. Tail conico-cylindrical with slightly bulbous tip. Caudal glands entirely in tail.

Spicules elongate, 260 μm (4.5 a.b.d.) with narrow distal portion.

Gubernaculum well developed. Two opposed testes lying to the left of the gut.

Distribution. West Ireland (sand and shells at 44 m); Isles of Scilly (shell gravel at 52 m).

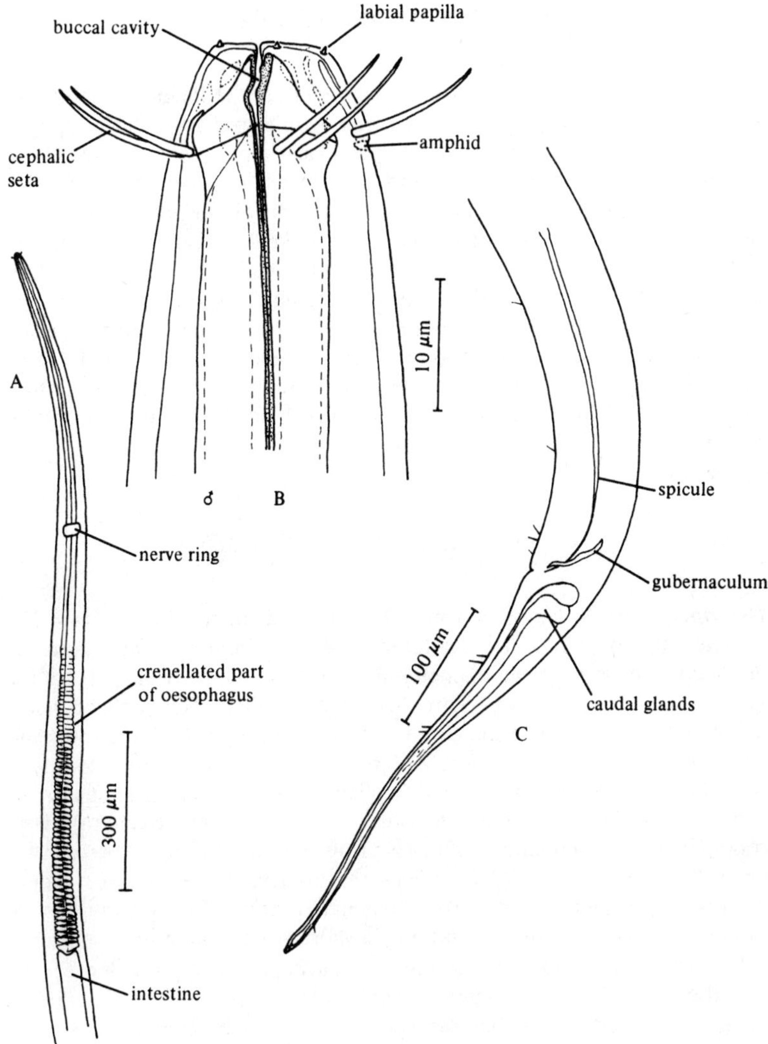

Fig. 67. *Crenopharynx marioni*. A, Anterior end; B, Male head (medial view); C, Male tail. Original.

Genus PHANODERMA Bastian, 1865

Key: Wieser, 1953

Phanoderma is distinguished from other genera of the Phanodermatidae by the strong trilobed cephalic capsule, often with a striated posterior margin, short spicules and the presence of a precloacal supplement. Wieser (1953) subdivided *Phanoderma* species into two subgenera according to whether ocelli are present or absent. However, the pigment in the ocelli can fade on preservation so these structures must be looked for carefully. Only five species thus far have been described without ocelli. Species are differentiated mainly by the shape of the tail, spicule size and structure of the cephalic capsule. In addition to the two species described below, a third species, *P. laticolle* (Marion, 1870) was reported by Southern (1914) and Moore (1971) from British waters but not described. However, its tail is longer than the other two British species according to the description of the male by De Man (1876).

Species : 31

Phanoderma albidum Bastian, 1865

(Fig. 68)

Description. Body length 4.5 mm. Maximum diameter 165 μm (a = 27). Cuticle smooth with a few scattered very short somatic setae. Anterior gradually attenuated; head diameter about 15% of the body diameter at the posterior end of the oesophagus. Six labial papillae. Ten cephalic setae; longer submedian setae 8 μm (0.5 h.d.), shorter 6.5 μm (0.4 h.d.). Ocellus 40 μm (2.6 h.d.) from anterior. Amphids situated just posterior to lateral cephalic setae. Buccal cavity small. Cephalic capsule present but not strongly cuticularised and extending anteriorly as three lobes. Excretory pore conspicuous, situated anterior to nerve ring at about 0.13 times oesophagus length. Nerve ring at 0.42 times oesophagus length. Oesophagus widens posteriorly and has the characteristic crenellated outline. Tail short and conico-cylindrical. Caudal glands extend only a short distance anterior to anus.

Overies paired, opposed and reflexed. Vulva at 58% of body length.

Although females are sometimes quite abundant, males of *P. albidum* have never been found, but the species seems to be characterised by a combination of the lack of longitudinal striations of the cephalic capsule, cephalic setae length, tail shape and position of ocelli and excretory pore.

Distribution. Falmouth (on weed in tide-pool); Plymouth (on seaweed); Northumberland (kelp holdfast); Isles of Scilly (seaweed and kelp holdfast).

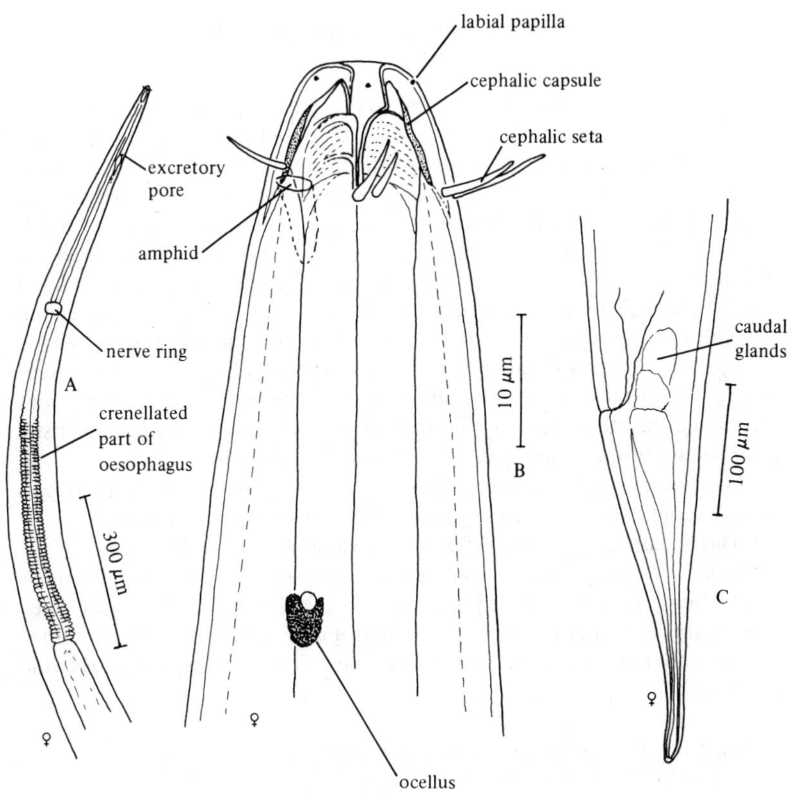

Fig. 68. *Phanoderma albidum*. A, Anterior end; B, Female head; C, Female tail. Original.

Phanoderma cocksi Bastian, 1869
(Fig. 69)

Description. Body length 3.6–6.4 mm. Maximum diameter 100–200 μm (a = 28–36). Cuticle smooth with scattered short somatic setae. Head small, the body width sharply decreasing between ocellus and cephalic setae; head diameter about 20% of body diameter at posterior end of oesophagus. Six labial papillae. Ten cephalic setae: longer submedian setae 15 μm (0.5 h.d.), shorter 13.5 μm (0.4 h.d.). Ocellus 50–55 μm (1.7 h.d.) from anterior. Amphids situated just posterior to lateral cephalic setae. Buccal cavity small but surrounded by three cuticular extensions of the cephalic capsule; the subventral lobes extend further anterior than the dorsal lobe. Cephalic capsule well developed and posteriorly bears longitudinal striations. Excretory pore conspicuous, situated anterior to nerve ring at about 0.05 times oesophagus length. Oesophagus widens posteriorly and has the characteristic crenellated outline. Tail short, conical with slightly bulbous tip.

Male has elongated spicules, 180–200 μm long and a cuticularised precloacal supplement situated level with the proximal end of spicules.

Ovaries paired, opposed, reflexed. Vulva at 60% of body length.

There are no specimens available of males of *P. cocksi* and Bastian's (1865) figure is rather inadequate now, so the figure here has been taken from Inglis' (1962) description of a male from the Mediterranean (described by him under the name *P. parafilipjevi*, now considered synonomous with *P. cocksi*).

Distribution. Falmouth (on weed in tide-pool).

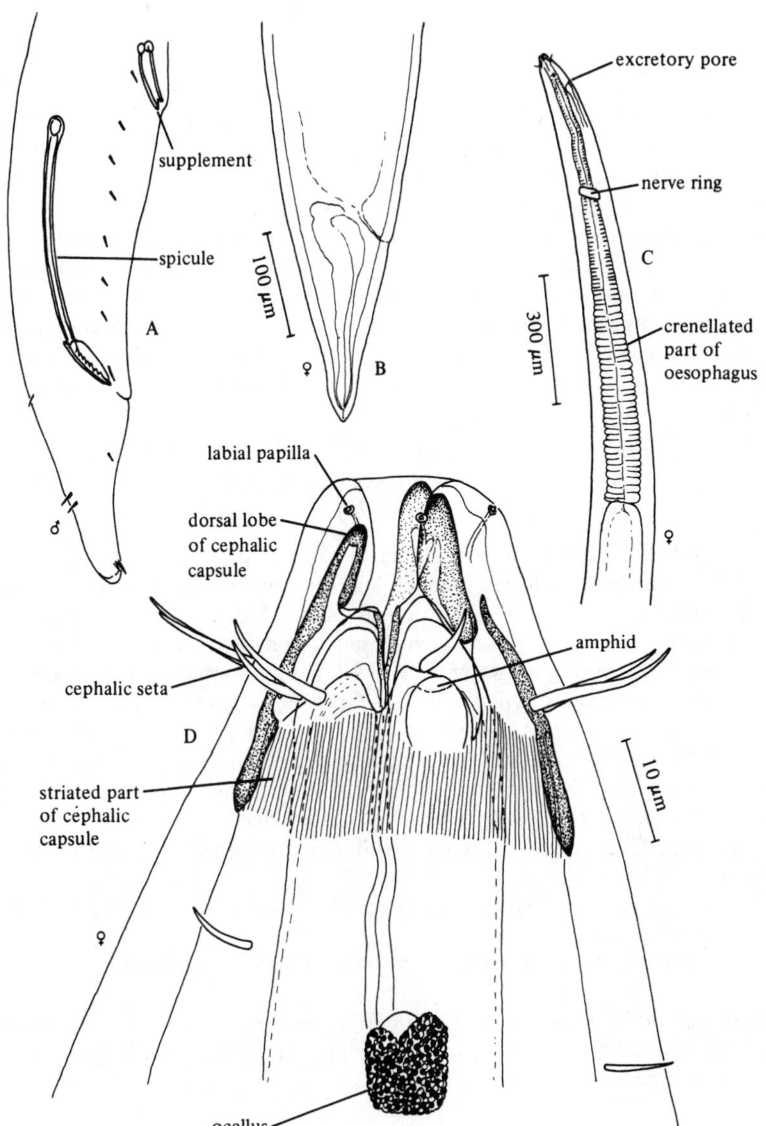

Fig. 69. *Phanoderma cocksi.* A, Male tail (from Inglis, 1962); B, Female tail; C, Anterior end; D, Female head. Original.

Genus ANTICOMA Bastian, 1865

Keys: Wieser, 1953; Wieser and Hopper, 1967
Anticoma species have a small conical buccal cavity, characteristic rows of lateral cervical setae, and a tubular precloacal supplement in the male. There are two main groups of species; those with the excretory pore anterior to or level with the cervical setae (exemplified by *A. acuminata*) and those with the excretory pore posterior to the cervical setae (e.g. *A. eberthi*). A third species, *A. pellucida* Bastian, 1865, has been recorded from the British Isles but the discussion is still open as to whether this species can be distinguished from *A. acuminata* and we are unable to give any characters which unequivocally separate the two. Species within these two main groupings are distinguished on rather fine points such as the exact position of the excretory pore, shape of tail, length of cephalic setae, and male copulatory apparatus.

Species : 42

Anticoma acuminata (Eberth, 1863)
(Fig. 70)

Description. Body length 1.9–2.1 mm. Maximum diameter 57–69 μm (a = 30–35). Cuticle smooth. Six rounded lips. Six small rounded labial papillae. Six longer cephalic setae 7–9 μm (up to 0.7 h.d.), shorter four about two-thirds this length. Buccal cavity, small, conical. Amphids pocket-like with elliptical openings 3–4 μm wide (0.2 times c.d.) in males, a little less in females. Oesophagus cylindrical, 0.2 times body length. Excretory pore about 2 times h.d. from anterior in male, just over 1 h.d. in female. Nerve ring half-way down oesophagus length. 4–5 cervical setae in single lateral files, anterior 2.8–3.5 h.d. from anterior. Tail with slightly more than distal half filiform, 6.2 a.b.d. long. Caudal glands confined to tail region.

Spicules 47–55 μm (arc), arcuate with a central lamella for most of their length.

Gubernaculum 13–19 μm, swollen distally. Tubular supplement 1.5 a.b.d. in front of cloaca.

Ovaries paired, equal, reflexed. Vulva at 47–49% of body length.

Distribution. Falmouth; Plymouth (intertidal seaweeds); North East England (*Laminaria* holdfasts); Exe estuary (intertidal sand); Isles of Scilly (intertidal sand, holdfasts, seaweeds).

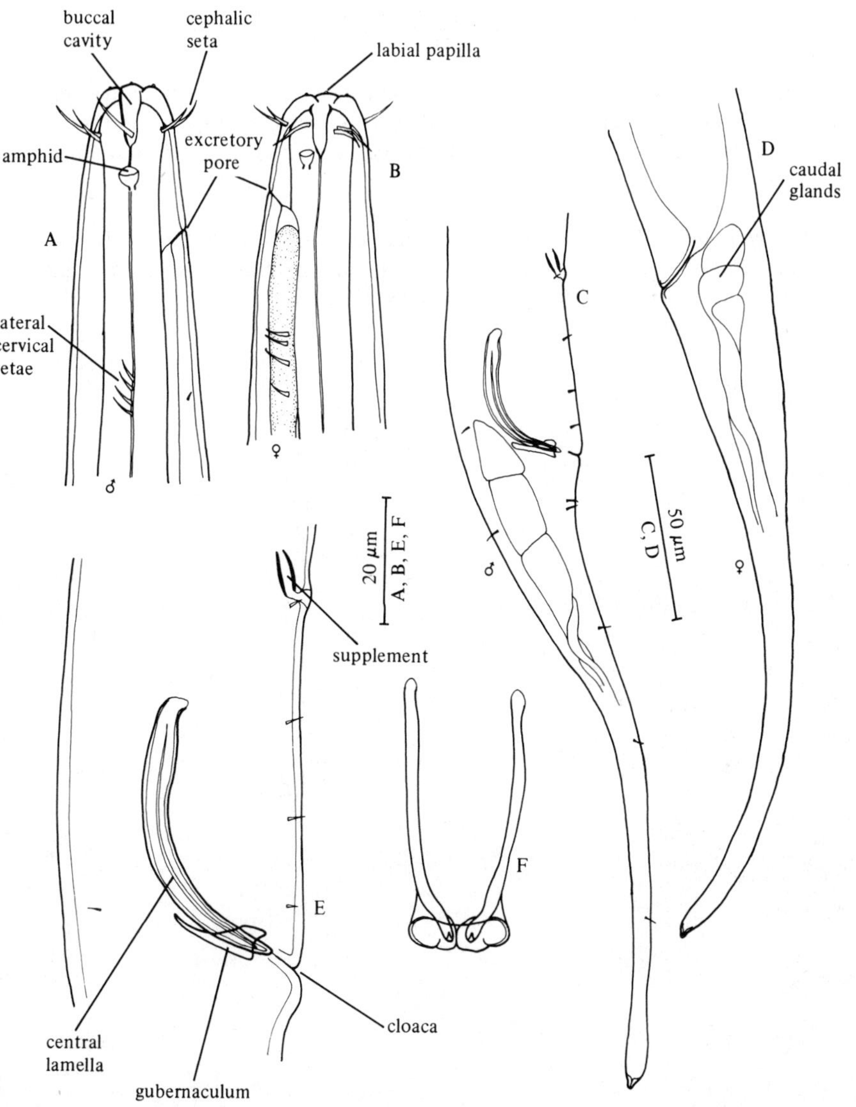

Fig. 70. *Anticoma acuminata*. A, Male head; B, Female head; C, Male tail; D, Female tail; E, Lateral view of male cloacal region; F, Ventral view of spicules and gubernaculum. Original.

Anticoma eberthi Bastian, 1865
(Fig. 71)

Description. Much larger than preceding species: body length 6–7 mm (a = 55–60). Cephalic setae 25 μm. Cervical setae commence 1.5 h.d. from anterior. Excretory pore well posterior to cervical setae. Tail 3–4 a.b.d. long. Spicules 1.3 a.b.d. long. Supplement 1.5 a.b.d. in front of cloaca.

Distribution. Falmouth (*Corallina* in tide-pool); Clare Island, West Ireland (intertidal); North East England (*Laminaria* holdfasts); Isles of Scilly (intertidal *Corrallina*, subtidal sand at 18–21 m).

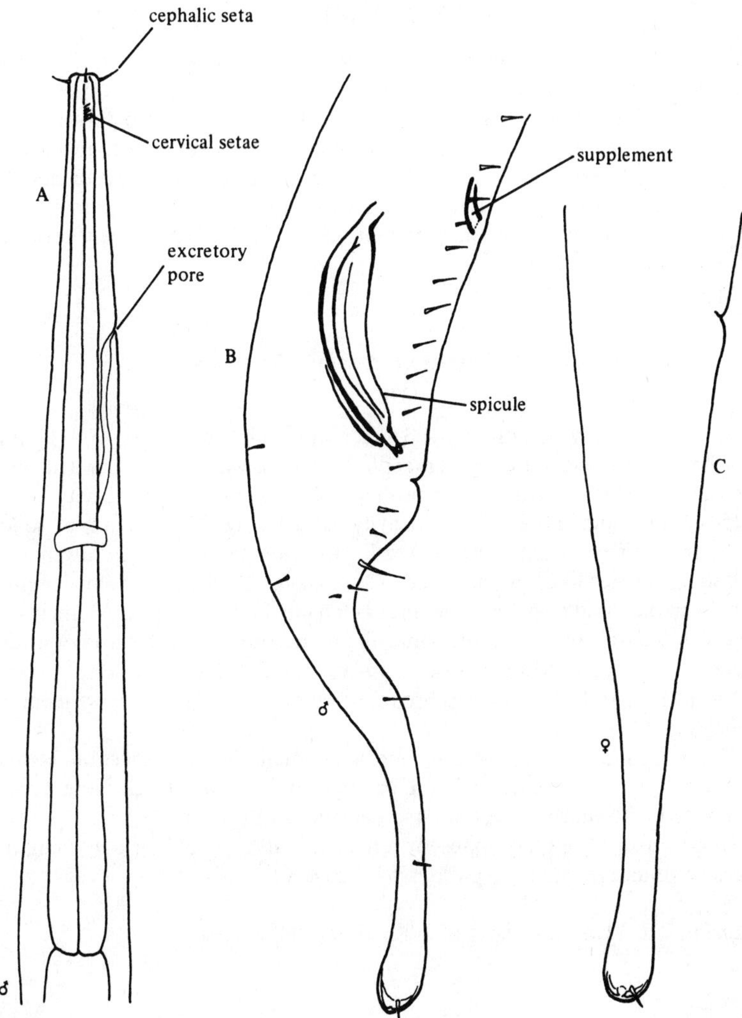

Fig. 71. *Anticoma eberthi*. A, Anterior end (from Bastian, 1865); B, Male tail; C, Female tail (from De Man, 1889).

Genus DOLICHOLAIMUS De Man, 1888

Dolicholaimus is distinguished from the other members of the Ironidae included here (*Thalassironus* and *Trissonchulus*) in having only three single teeth in the buccal cavity (in the other two genera, there are two dorsal teeth, sometimes fused at their bases, and two single subventral teeth); also in the shape of the tail which has a distal cylindrical section. The single or double nature of the dorsal tooth in the Ironidae can usually be seen best in *en face* preparations, but can often be seen more clearly in the replacement teeth of juveniles (see Fig. 75A).

Species: 1

Dolicholaimus marioni De Man, 1888
(Fig. 72)

Description. Body length 4.1–4.3 mm. Maximum diameter 107–118 μm (a = 36–38). Cuticle smooth. Ten cephalic papillae small and rounded, the submedian pairs scarcely distinguishable as double. Buccal cavity long (115–118 μm) and tubular, walls heavily cuticularised. Three anterior solid teeth rather small; replacement teeth posterior to this set in juveniles. Amphids pocket-like, 10 μm wide. Oesophagus 0.15 times body length, narrow in the middle with a pronounced elongate bulb at its base, 63 μm wide and 133–145 μm long. Tail the same shape in both sexes, distal two-thirds tapering, proximal third cylindrical, tip swollen, 2.4–2.9 a.b.d. long.

Spicules paired, 76 μm long, broad, squared proximal ends, two median lamellae.

Gubernaculum paired, 45 μm long with small caudal projection in the middle. 6–7 small conical ventral papillae in front of cloaca, sometimes associated with more or less marked elevations of the cuticle.

Ovaries paired, equal, opposed, reflexed. Vulva at 53% of body length. Eggs large and elongate, typically 254 × 80 μm.

Distribution. Plymouth; Isles of Scilly (intertidal algae).

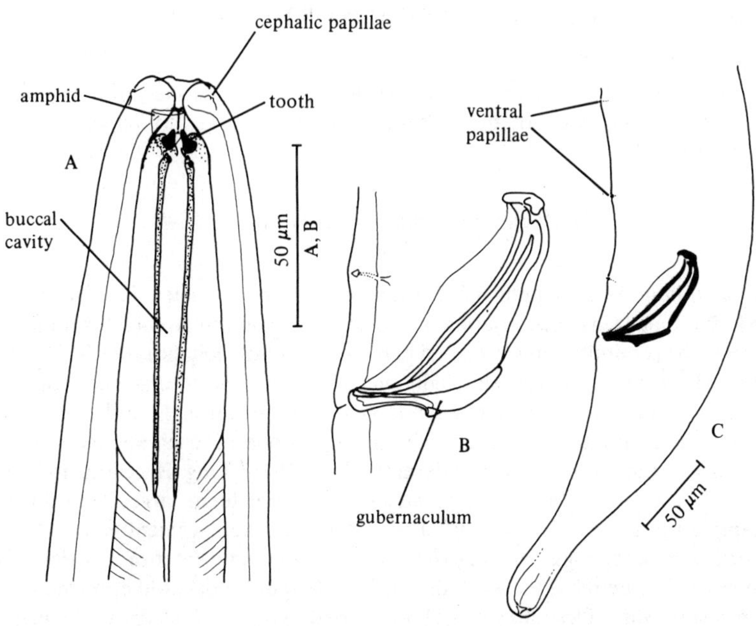

Fig. 72. *Dolicholaimus marioni*. A, Male head; B, Spicules and gubernaculum; C, Male tail. Original.

Genus THALASSIRONUS De Man, 1889

Key: Vitiello, 1970

Thalassironus has a double dorsal tooth and is distinguished from the next genus, *Trissonchulus*, by the presence of cephalic setae, not papillae.

Species : 4

Thalassironus britannicus De Man, 1889
(Fig. 73)

Description. Body length 6.5–7.5 mm. Maximum diameter 73–85 μm (a = 84–88). Fine transverse striation of cuticle sometimes discernible. Six rounded lips with six minute rounded labial papillae. Six long cephalic setae 24–28 μm, four shorter setae 10–12 μm. Eight characteristic cervical setae each consisting of a pair of stout setae closely applied together throughout their length; three on each lateral surface of the head and one on each median surface level with the posterior of the lateral triad; 14–19 μm long. Buccal cavity long (53–77 μm) and tubular. Paired single subventral teeth larger than dorsal double-tipped one. Larvae with a posterior set of replacement teeth. Outer surface of oesophagus cuticularised into a weak capsule anteriorly, acting as a support for the basal plates of the onchia. Amphids pocket-like, opening by crescentic slits. Oesophagus 0.11 times body length, broadening gradually towards its base. Tail tapering throughout its length in both sexes, 2.7–2.8 a.b.d. long; paired terminal setae 6–9 μm long.

Spicules paired, 62–72 μm long, cephalate proximally with a short median list near the proximal end. Gubernaculum paired, 30–38 μm long, expanded distally into concave cup-like swellings which are heavily cuticularised. A single median precloacal papilla and 8–9 pairs of small rounded papillae in two subventral files down the length of the male tail.

Ovaries paired, equal, opposed, reflexed. Vulva at 61% of body length. Eggs elongate, typically 71 × 32 μm.

Distribution. Penzance; Isles of Scilly; West Ireland (intertidal and subtidal sand).

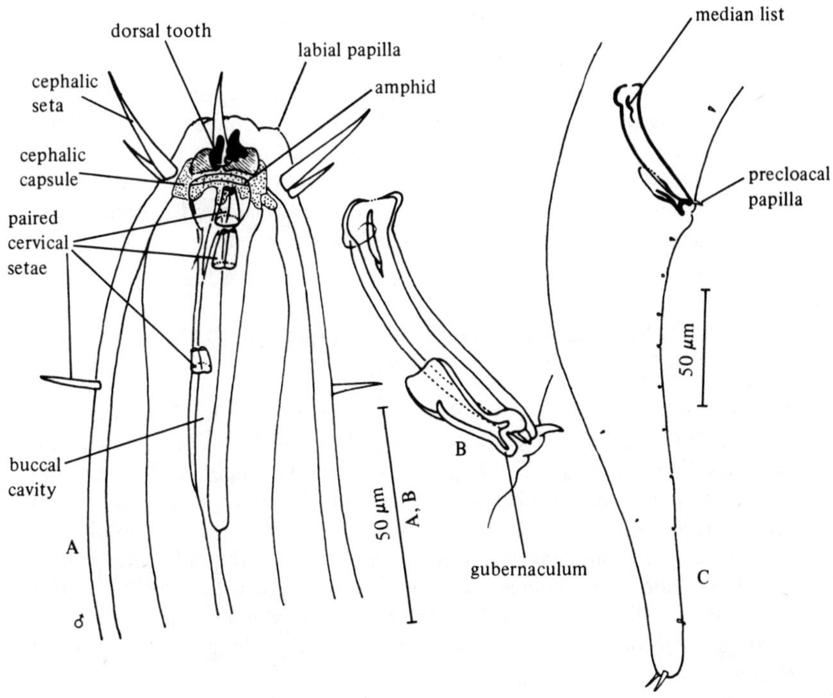

Fig. 73. *Thalassironus britannicus*. A, Male head; B, Spicules and gubernaculum; C, Male tail. Original.

Genus TRISSONCHULUS Cobb, 1920

Trissonchulus species have (so far as is known) four teeth in the buccal cavity, with the dorsal pair fused together to varying degrees, often appearing completely separate in the replacement teeth of the juveniles (Fig. 75A). They have ten cephalic papillae and relatively short tails (with no cylindrical section distally). *Trissonchulus benepapillosus* is distinct from the other two British species in having a conical tail, whereas *T. obtusus* and *T. oceanus* can easily be separated on the basis of the papillate or non-papillate anterior end.

Species : 10

Trissonchulus benepapillosus (Schulz, 1935)
(Fig. 74)

Description. Body length 2.4–2.9 mm. Maximum diameter 53–66 μm (a = 40–48). Cuticle smooth. Head dome-shaped, set off by a deep constriction. Six small labial papillae. Ten large conical cephalic papillae. Buccal cavity lined with many small cuticular denticles, which occupy a circular area in lateral view. Solid curved teeth, the dorsal one comprising two fused teeth (double-tipped), subventral teeth single, larger. Amphids cup-shaped, 12–14 μm wide. Oesophagus 0.15 times body length with distinct anterior peribuccal swelling; posterior third enlarged but no definite bulb. Nerve ring at about 50% of oesophagus length. Tail 2.2 a.b.d. long, conical. Caudal glands open terminally.

Spicules 46–53 μm long, very wide and blunt proximally, with three central lamellae.

Gubernaculum 12–13 μm, paired, with swellings of a complex folded structure distally. No precloacal supplements.

Ovaries paired, equal, opposed, reflexed. Vulva at 52–54% of body length.

Distribution. Exe estuary (intertidal sand).

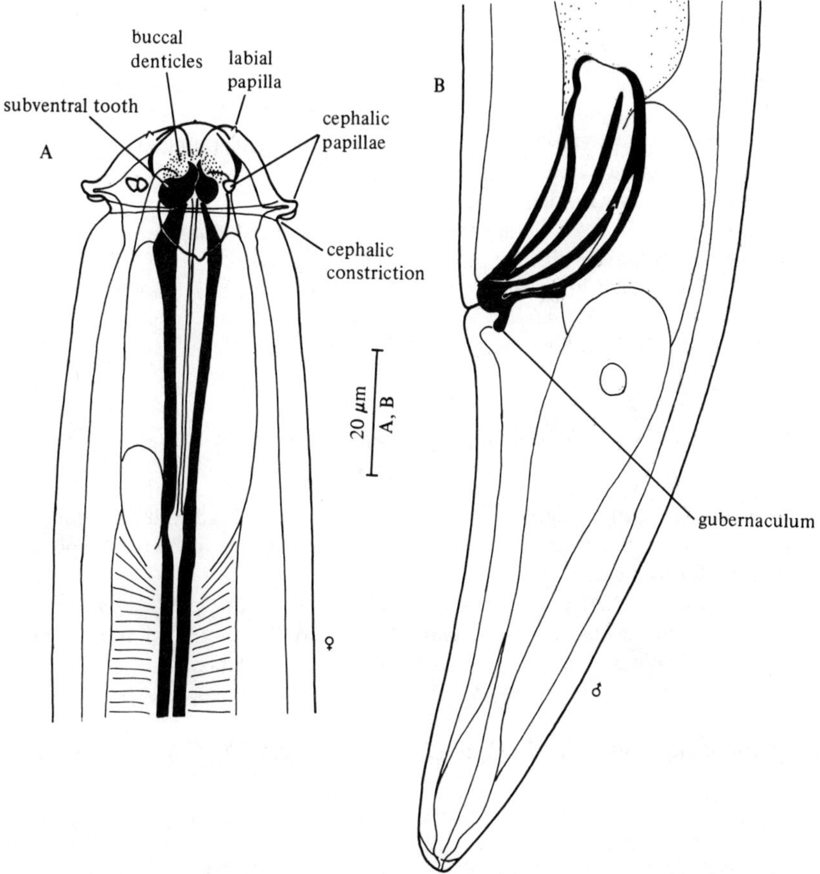

Fig. 74. *Trissonchulus benepapillosus*. A, Female head; B, Male tail. Original.

Trissonchulus obtusus (Bresslau and Stekhoven, 1935)
(Fig. 75A, B, C)

Description. Body length 4.0–4.5 mm (a = 64–65). Head and buccal cavity similar to *T. benepapillosus*, but with rows of conical papillae on anterior end of body. Amphids cup-like, 0.26 times c.d. wide. Tail rounded, 1.2 a.b.d. long, caudal glands open ventrally, subterminal.

Spicules 1.2 a.b.d. long.

Gubernaculum 0.43 a.b.d., paired, swollen at distal ends. 4–5 precloacal supplements. Male with four pairs of papillae on tail.

Vulva at 49% of body length, female tail with two apical papillae.

Distribution. Isles of Scilly (*Laminaria* holdfasts).

Trissonchulus oceanus Cobb, 1920
(Fig. 75D, E)

Description. Body length 3.2–5.6 mm (a = 42–56). Head and tail similar to *T. obtusus*, but no rows of papillae on anterior end of body. Amphids 0.29–0.35 times c.d. wide.

Spicules 60 μm long. Gubernaculum 27 μm. A swelling pierced by a pore on the anterior lip of the cloaca, and a similar swelling on the tail behind the exit of the caudal glands. 9 precloacal pore-like supplements.

A single post-vulvar ovary. Vulva at 41–3% of body length.

Distribution. North East England (*Laminaria* holdfasts).

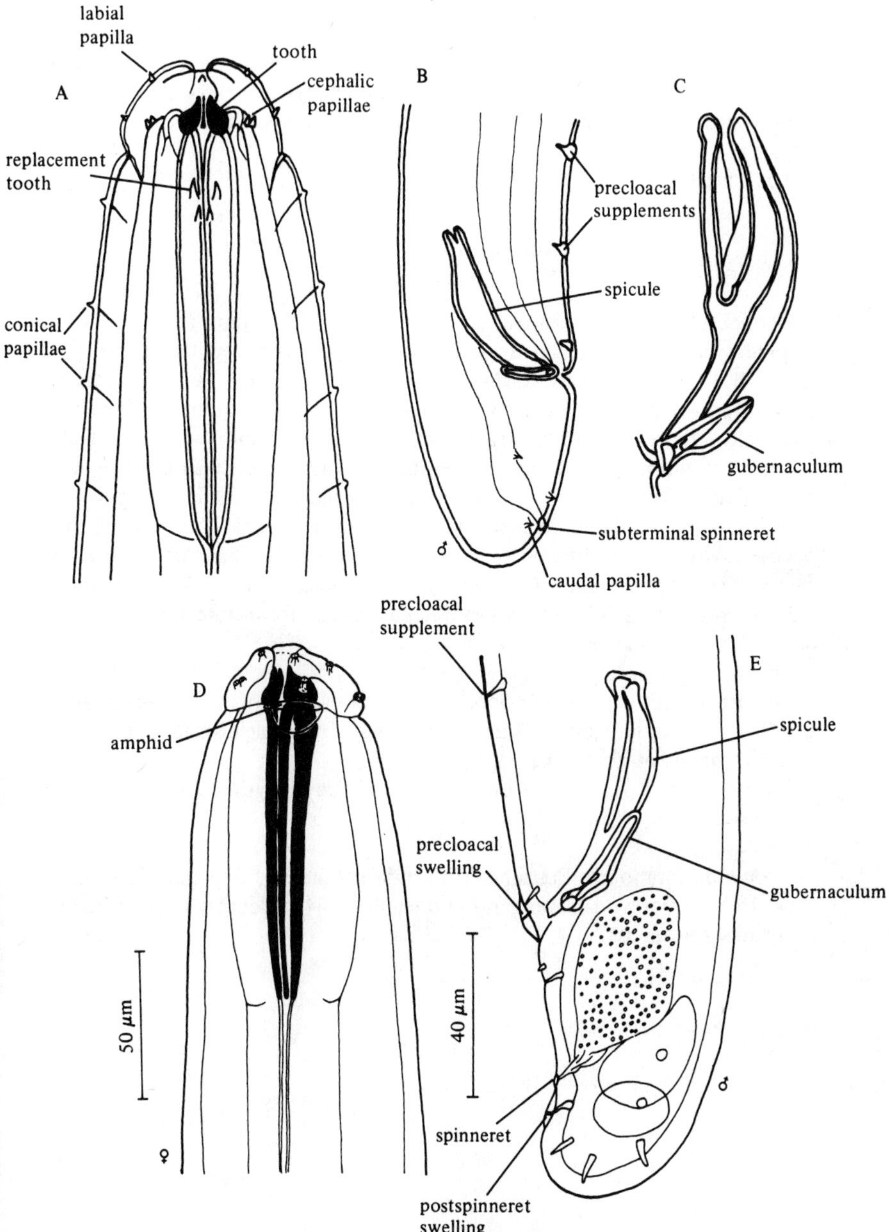

Fig. 75. *Trissonchulus obtusus*. A, Head of young specimen showing replacement teeth; B, Male tail; C, Spicules and gubernaculum (from Stekhoven, 1935). *Trissonchulus oceanus*, D, Female head (from Stekhoven, 1943); E, Male tail (from Gerlach, 1967).

Genus CYLICOLAIMUS De Man, 1889

Cylicolaimus species have a large tubular buccal cavity with a complicated array of teeth and tooth-like structures. The males have rows of precloacal spines and a gubernaculum with a strongly developed dorso-caudal apophysis.

Species : 3

Cylicolaimus magnus (Villot, 1875)
(Fig. 76)

Description. Very large species: body length 21–34 mm (a = 100–170). Cephalic capsule with lobed posterior margin. Ten cephalic setae 0.7 times h.d. Amphids pocket-like, immediately behind lateral cephalic setae. Buccal cavity with a circle of small pointed teeth at the anterior end, and below these are three larger teeth (one dorsal and two subventral) attached to the inner surfaces of the lips. On the lateral sides of the inner walls of the buccal cavity are two triangular plates, with sharply pointed anterior corners, resembling teeth but lying flat against the wall of the buccal cavity. Three cuticularised rounded teeth at the base of the buccal cavity. Six files of setae in cervical region. Oesophagus 0.14 times body length. Nerve ring 0.17 times oesophagus length from anterior. Tail just under 2 a.b.d. long, bluntly rounded.

Spicules 230 μm long.

Gubernaculum with strong dorso-caudal apophysis. A median precloacal papilla 0.6 a.b.d. in front of cloaca. Four rows of precloacal setae; outer rows of 10–11, inner of about 30.

Ovaries paired, reflexed. Vulva at 61% of body length. Female tail with sparse setae.

Distribution. Falmouth; Penzance (intertidal); Blacksod Bay and Clew Bay, West Ireland (intertidal and sand and shells at 44 m depth); Isles of Scilly (intertidal seaweeds).

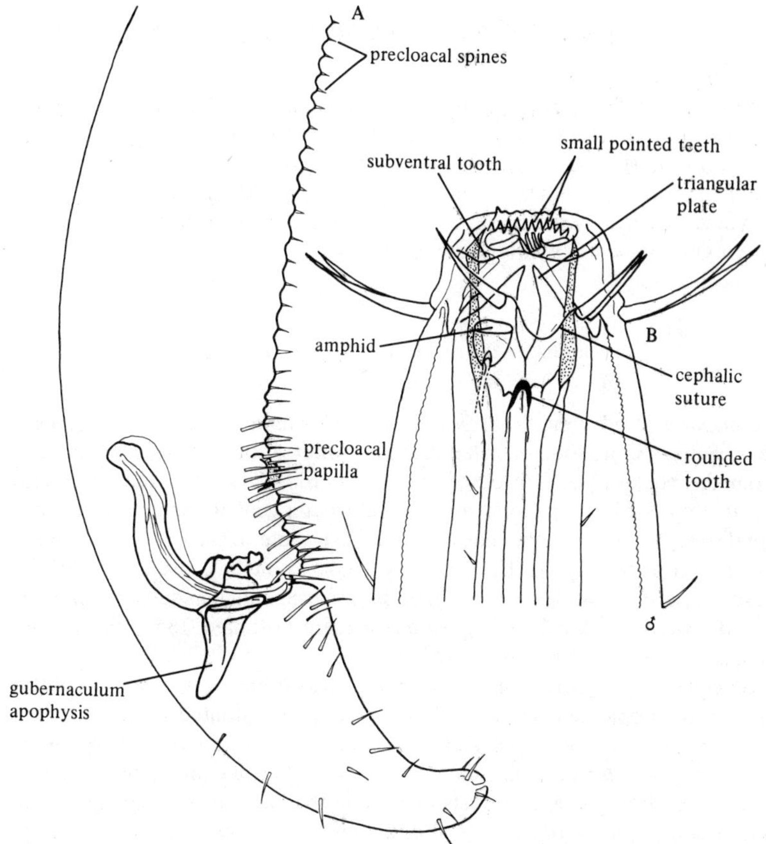

Fig. 76. *Cylicolaimus magnus*. A, Male tail; B, Head (from Southern, 1914).

Genus LEPTOSOMATIDES Filipjev, 1918

Key: Platonova, 1976

The species in this genus mainly vary in the number and arrangement of the precloacal subventral papillae and the shape of the gubernaculum. However, most of the species have been poorly described. The genus itself is distinguished from other similar genera with relatively weakly developed cephalic capsules by the presence of the precloacal supplement and papillae, and a gubernaculum with a dorsal apophysis.

Species : 10

Leptosomatides euxinus Filipjev, 1918
(Fig. 77)

Description. Body length 11.9–15.2 mm. Maximum diameter 170–225 μm (a = 68–69). Short somatic setae present. Anterior of body tapers gradually from the posterior end of oesophagus (c.d. 160–170 μm) to the head (c.d. 47–48 μm). Six labial papillae and twelve 5–6 μm cephalic setae. Six longitudinal rows of cervical setae plus a group of extra lateral cervical setae situated just posterior to amphid. Buccal cavity small. Cephalic capsule present; the posterior margin can be seen as an undulating subcuticular suture. Amphids 8–9 μm wide (0.15–0.20 c.d.). Ocelli present, situated 135–140 μm from anterior. Tail very short, rounded.

Spicules 160–165 μm. Gubernaculum has dorsal apophysis. Ventral precloacal supplement present. Subventral spines situated between cloaca and supplement. Seven precloacal subventral papillae; the first two are level with and just anterior to the supplement, the other five are further anterior (Fig. 41B). Each papilla consists of two short longitudinal ridges of cuticle with a small spine in the middle. Two opposed, outstretched testes; anterior situated to the left, the posterior to the right of the gut.

Two opposed, reflexed ovaries.

Distribution. Isles of Scilly (kelp holdfasts).

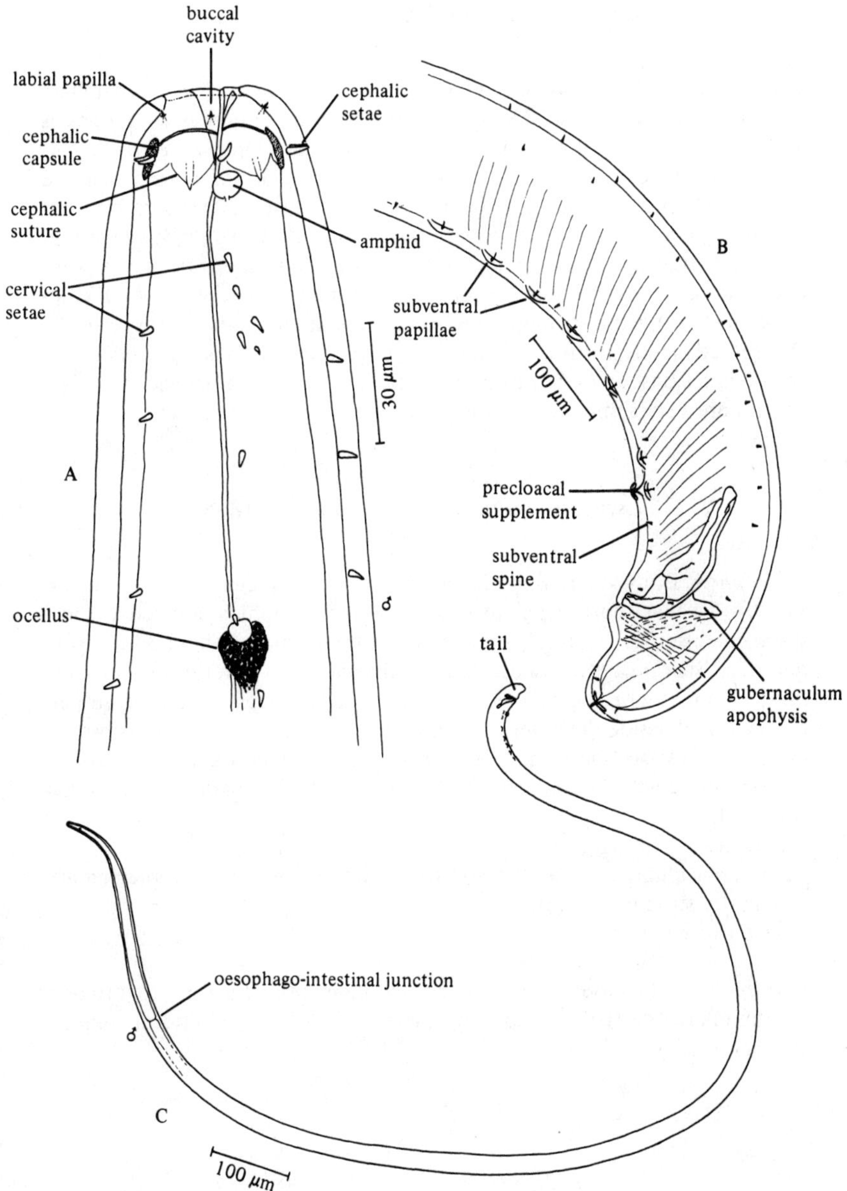

Fig. 77. *Leptosomatides euxinus*. A, Male head; B, Male tail; C, Entire male. Original.

Genus LEPTOSOMATUM Bastian, 1865

Key: Platonova, 1976

Leptosomatum has a weak cephalic capsule, no buccal cavity, papillate cephalic setae, and ocelli. In the male the gubernaculum has no apophysis and there is no precloacal supplement or papillae. Only *L. elongatum* is recorded from Britain, and this is poorly described. Bastian (1865) and De Man (1893) provide the only descriptions of British material, both rather primitive. The data given below are taken from Bastian's work whilst Fig. 78A is a drawing of a juvenile from Norway; no other specimens were available for study. Wieser (1951) reported another species, *L. bacillatum* (Eberth, 1863), from Plymouth but since there are no figures nor specimens the record needs further substantiation. A third species *L. gracile* was recorded by Bastian (1865) from the same locality as *L. elongatum*; the two species are very similar and are here considered to be the same.

Species : 27

Leptosomatum elongatum Bastian, 1865
(Fig. 78)

Description. Body length 8.3–12.7 mm Maximum diameter 127–191 μm (a = 60–80). Twelve short papillate cephalic setae. Amphids situated a short distance posterior to lateral cephalic seta. Ocelli situated about 3 h.d. from anterior. Buccal cavity absent, but dorsal segment of anterior oesophageal lumen lining is more heavily cuticularised. Cephalic capsule present, resembling a shield beside the anterior part of the oesophagus and most easily seen in optical cross-section (Fig. 42A). Nerve ring situated at about 25% of the oesophagus length. Oesophagus 0.11–0.14 times body length. Tail short and rounded.

Spicules 75–100 μm.

Gubernaculum lies parallel to distal end of spicules. Supplementary copulatory structures absent.

Two ovaries, opposed.

Distribution. Falmouth (in sponge); Plymouth (among hydroids); Northumberland (kelp holdfasts); Isles of Scilly (sublittoral among sponges and hydroids).

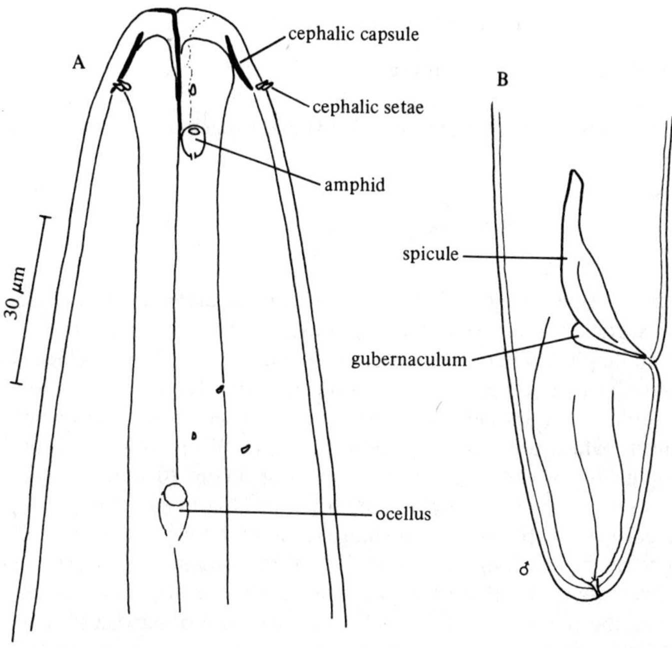

Fig. 78. *Leptosomatum elongatum*. A, Juvenile head (original); B, Male tail (from Bastian, 1865).

Genus PLATYCOMA Cobb, 1894

The small buccal cavity with three small teeth, and the paired stout amphidial setae in the male characterise this genus. The following description is based on a re-examination of Southern's (1914) male specimen from Clew Bay.

Species : 1

Platycoma cephalata Cobb, 1894
(Fig. 79)

Description. Body length 8.8–10.7 mm. Maximum diameter 90–120 μm (a = 89–98). Cuticle smooth except for the constriction zone between the cephalic setae and amphid, where the cuticle bears transverse striations which can be seen most easily in optical cross-section (Fig. 79B). Buccal cavity absent but the oesophagus bears anteriorly three cuticularised teeth (one dorsal, two subventral) which protrude beyond the lips. Six labial papillae, their position identifiable by subcuticular innervation. Ten 30 μm (0.7 h.d.) long stout cephalic setae. Amphids 38 μm from anterior, 15 μm (0.3 c.d.) wide. The cuticle posterior to the amphid opening carries two stout setae (Fig. 79B), the dorsal being slightly longer (27 μm), so that the amphid opening lies behind these structures. There are somatic setae present in the area between the head and the nerve ring (Fig. 79A) but these are otherwise absent. The oesophagus is narrow in the anterior part, but is wider in the posterior 40%. The tail is conico-cylindrical.

Spicule 80 μm long with complicated lateral structures distally (Fig. 79D).

Gubernaculum has paired dorsal apophyses. Paired subventral setae present posterior to and just anterior to cloaca. Ventral precloacal supplement situated 95 μm anterior to cloaca and, further anteriorly, a pair of subventral spines situated on papillae.

Vulva situated nearly 66% of the body length from the anterior.

Distribution. West Ireland (sand and shells at 44 m).

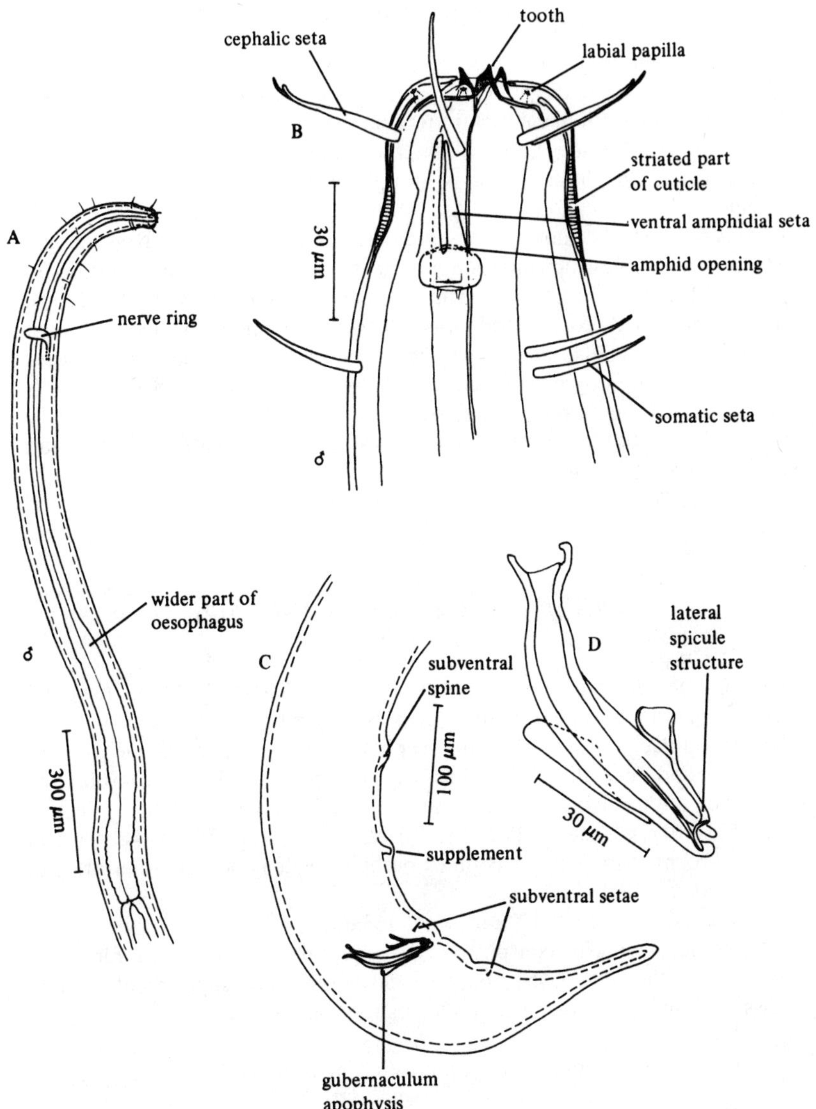

Fig. 79. *Platycoma cephalata*. A, Anterior end of male; B, Male head; C, Male tail; D, Spicules and gubernaculum. Original.

Genus PSEUDOCELLA Filipjev, 1927

Key: Platonova, 1976

Pseudocella has a strongly developed cephalic capsule which usually has a hollow ventral tooth-like structure termed the *tropis*. It is distinguished from *Thoracostoma* by the absence of true ocelli (i.e. with a lens) and a gubernacular apophysis which is at right angles to (not parallel to) the spicules. *P. trichodes* is the most frequently reported of the *Pseudocella* species but there has not been a good description since that of Bresslau and Stekhoven (1940), which differs somewhat from that of Stekhoven and Adam (1931). However, the characteristic feature seems to be the ventral structure near the tail tip. According to Filipjev (1927), the number and distribution of the setae and papillae around the cloaca can vary. The only other species recorded from European waters is *P. elegans* (Ditlevsen, 1926) which is described without papillae anterior to the supplement. However, it is possible that the author overlooked these structures and *P. coecum* and *P. elegans* are the same.

Species: 27

Pseudocella coecum (Ssaweljev, 1912)
(Fig. 80A, B)

Description. Body length 9.0 mm. Maximum diameter 220 μm (a = 41). Cuticle smooth; minute somatic setae present throughout the body. Six small labial papillae and ten 9–11 μm cephalic setae. Cervical setae present laterally and medially. Head bears a conspicuous cephalic capsule, the posterior suture lying posterior to the amphid. Ventral tropis present. Amphids 10 μm wide (0.25 c.d.). Buccal cavity small and without teeth. Ocelli absent. Nerve ring 0.3 times oesophagus length from anterior. Tail short and rounded.

Spicules unequal; left 190 μm, right 165 μm.

Gubernaculum with conspicuous dorsal apophysis. There is a file of 14 long setae either side of the cloaca, a ventral precloacal supplement and two rows of 7 subventral papillae. Testes paired, opposed.

Distribution. Northumberland (sublittoral fine sediment).

Pseudocella trichodes (Leuckart, 1849)
(Fig. 80C)

Description. There are several records of this species from Britain but no description. The following information and the figure comes from Stekhoven and Adam (1931), collected on the Belgian coast. Head basically similar to *P. coecum*. Tail bears a postcloacal supplement, situated near the tail tip,

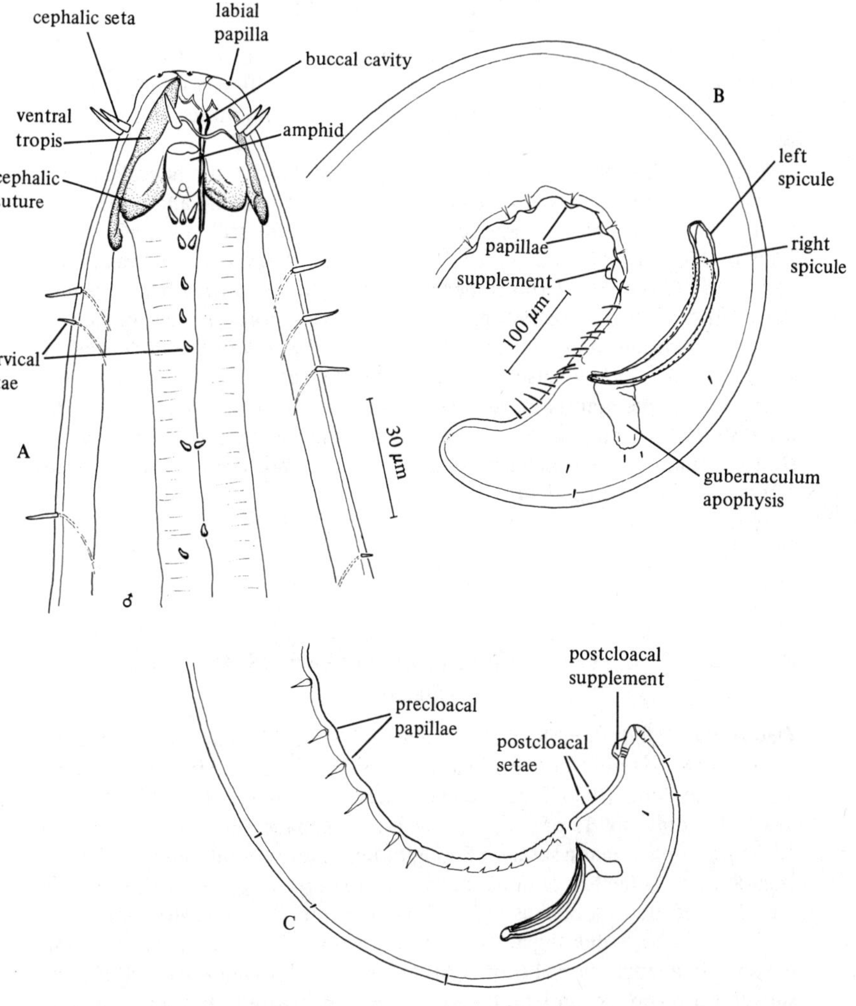

Fig. 80. *Pseudocella coecum*. A, Male head; B, Male tail (original). *Pseudocella trichodes*. C, Male tail (from Stekhoven and Adam, 1931).

with 4 grooves leading to the interior. Two long setae are situated posterior to cloca. Anterior to the cloaca are 15 papillae with small spines.

Distribution. West Ireland (under stones); Plymouth (among seaweed); Northumberland (in intertidal rock crevices and kelp holdfasts); Isles of Scilly (among seaweed).

Genus SYNONCHUS Cobb, 1894

Platonova (1970) erected a new genus *Eusynonchus* for *Synonchus brevisetosus* (Southern, 1914) based on supposed differences in dentition. However, since we failed to detect any major discontinuities in this regard, we here consider the genera to be identical.

Synonchus has one dorsal and two subventral teeth in the buccal cavity, a precloacal supplement and two subventral rows of papillae and setae in the male, and a clavate tail. Species in this genus are still relatively poorly known. The only previous description of a male *S. fasciculatus* was that of Cobb 1894 (from the Mediterranean) and the other two British species have not been described since 1940. *S. longisetosus* is a more slender animal than the other two and its precloacal supplement is level with the middle of the spicule. *S. fasciculatus* also has fairly cephalic setae but the tail is not so short, the supplement is more anterior and the subventral spines and setae are differently arranged. *S. brevisetosus* has the shortest cephalic setae and shorter somatic setae which are not as conspicuous posterior to the nerve ring as they are in *S. fasciculatus*.

Species: 11

Synonchus fasciculatus Cobb, 1894
(Fig. 81)

Description. Body length, male 10.8 mm, female 13.0 mm. Maximum diameter, male 164 μm, female 180 μm (a = 66.3 and 72.3). Six labial papillae. Ten 19 μm long (0.36 h.d.) cephalic setae in female. In the only male available, only six 15 μm long (0.29 h.d.) cephalic setae were observed; 18–25 μm long somatic setae in 8 longitudinal files present anterior to nerve ring. Somatic setae less abundant posterior to nerve ring. Nerve ring at about one-third of the oesophagus length from anterior. Buccal cavity narrow and conical, anteriorly bearing a pair of cuticularised subventral onchia and more posteriorly a single dorsal tooth. At its base, the buccal cavity wall has two subventral tooth-like indentations. The cephalic capsule has an undulating posterior suture. Amphids situated posterior and slightly dorsal to lateral cephalic seta; width 9–10 μm (0.16–0.18 c.d.). Oesophagus 0.17–0.18 times body length. Tail conico-cylindrical with bulbous tip.

Spicules 160 μm with faint ventral alae and a lateral spine at the tip. Gubernaculum has a prominent dorsal apophysis. Level with the proximal end of the spicule there is a ventral precloacal supplement. Further anteriorly, between 300–400 μm from the cloaca, there is a patch of about 14 anteriorly-facing spines, each situated on a small papilla. There are two subventral files of long setae situated between the cloaca and supplement.

Testes paired, opposed.

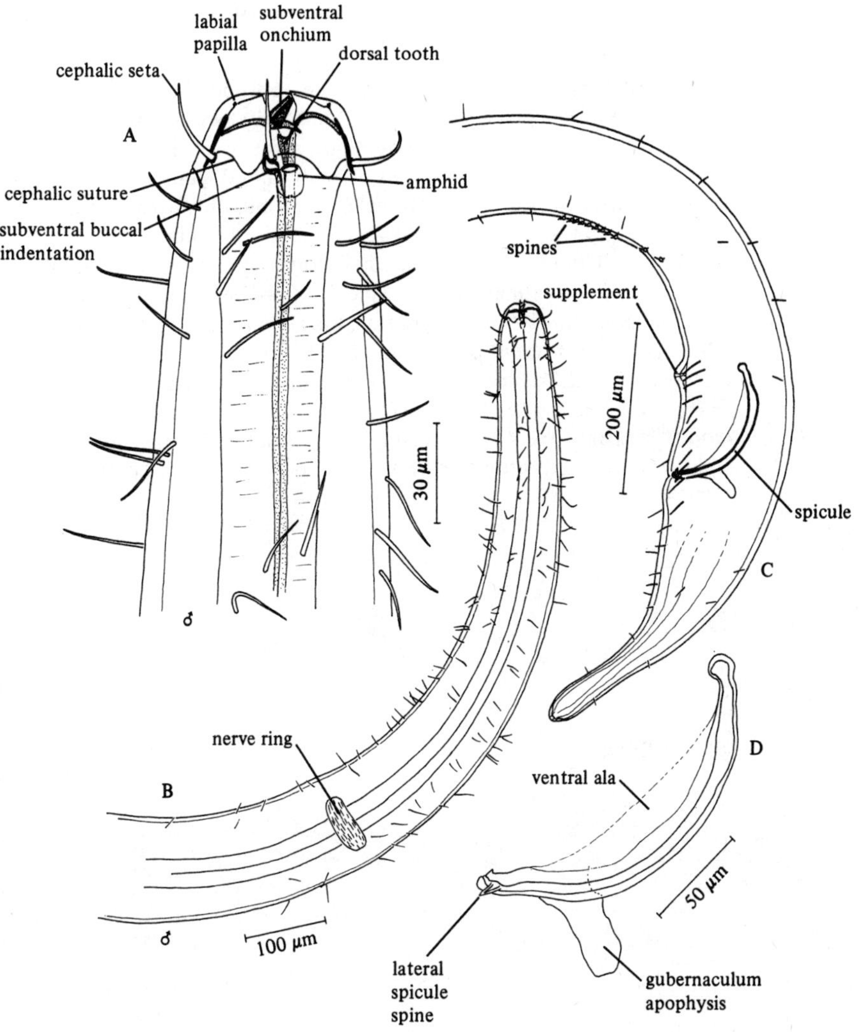

Fig. 81. *Synonchus fasciculatus*. A, Male head; B, Anterior end of male; C, Male tail; D, Spicules and gubernaculum. Original.

Two opposed, reflexed ovaries. Vulva at 49% of body length from anterior.

Distribution. Isles of Scilly (sublittoral among worm tubes, sand and gravel).

Synonchus brevisetosus (Southern, 1914)
(Fig. 82)

Description. Body length, male 13.4 mm, female 16.0 mm. Maximum diameter, male 290 μm, female 280 μm (a = 46.2 and 57.0). Six labial papillae. Ten cephalic setae about 0.15–0.2 h.d. long. Short somatic setae present only anterior to nerve ring. In female from Isles of Scilly lateral setae posterior to amphids appear flattened (Fig. 82A). Buccal cavity similar to *S. fasciculatus*. Amphids 0.13–0.20 times c.d. Tail conico-cylindrical with bulbous tip.

Spicules 168 μm. Gubernaculum with small dorsal apophysis. Precloacal supplement opposite proximal end of spicule. Two subventral files of precloacal setae; anterior 15 pairs are shorter, thicker and seated on rounded papillae. Setae are also present on the tail (Fig. 82D).

Two opposed, reflexed ovaries. Vulva at 59% of body length.

Distribution. Clew Bay, West Ireland (sublittoral, sand and shells); Isles of Scilly (sublittoral, coarse bottom); Anglesey (coarse shell gravel beach).

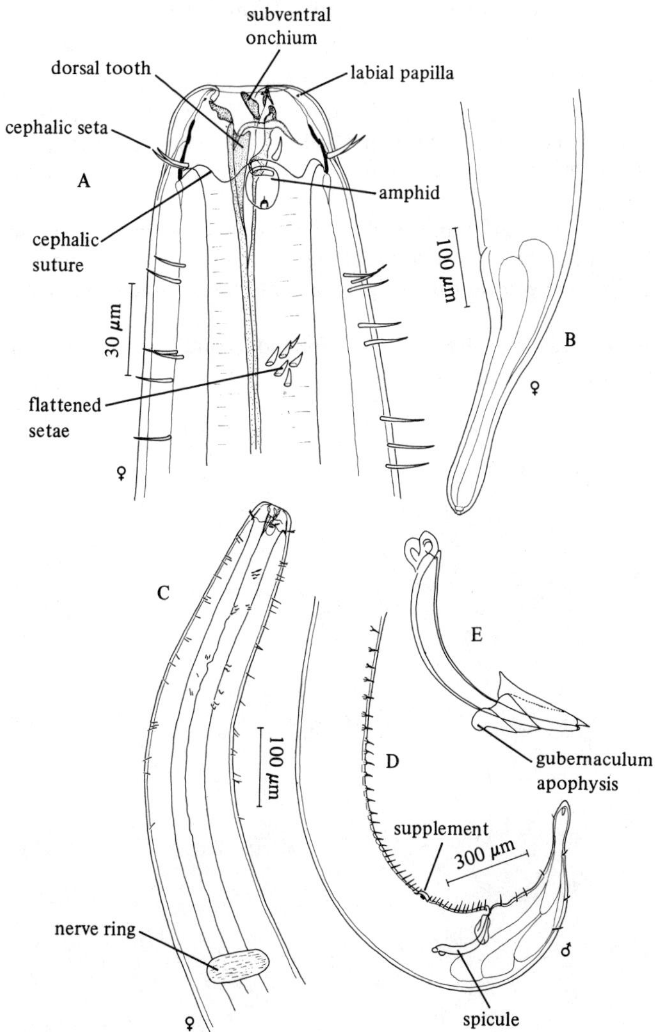

Fig. 82. *Synonchus brevisetosus*. A, Female head; B, Female tail; C, Anterior end of female (original). D, Male tail; E, Spicules and gubernaculum (from Southern, 1914).

Synonchus longisetosus (Southern, 1914)

(Fig. 83)

Description. Body length, male 17.1 mm. Maximum diameter 140 μm (a = 122). Ten cephalic setae, about 45 μm (0.5 h.d.). Numerous somatic setae arranged in six rows between head and nerve ring. Buccal cavity similar to *S. fasciculatus.* Amphids about 0.36 times c.d. Tail short, conical with bulbous tip.

Spicules about 180 μm. Precloacal ventral supplement level with the middle of the spicules. Two subventral files of setae posterior and anterior to cloaca. Two subventral rows of papillae anterior to supplement.

Distribution. Clew Bay, West Ireland (sublittoral, sand bottom); Northumberland (kelp holdfasts).

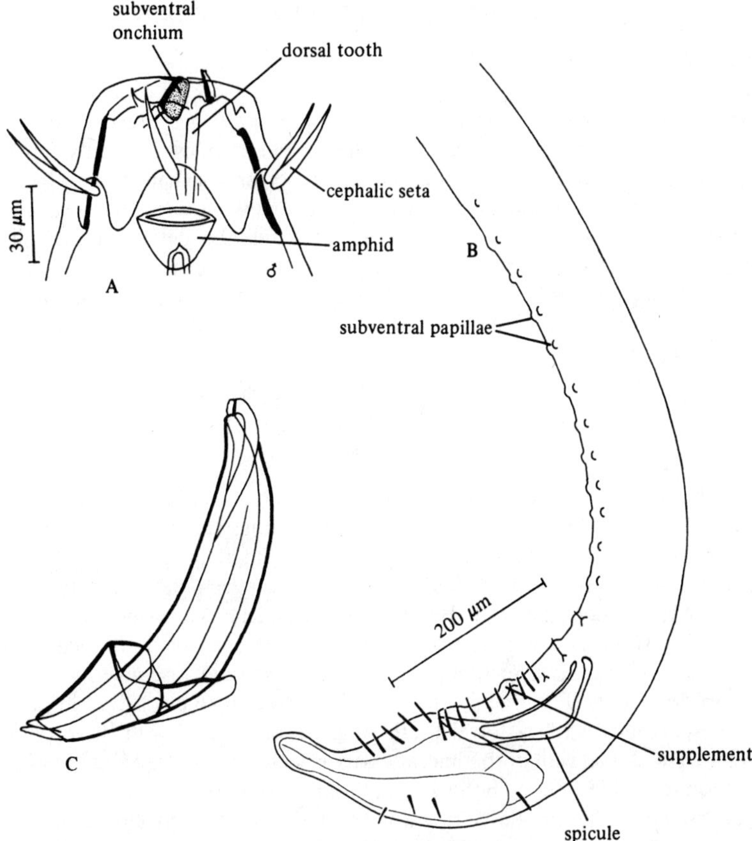

Fig. 83. *Synonchus longisetosus*. A, Male head; B, Male tail; C, Spicules and gubernaculum (from Southern, 1914).

Genus THORACOSTOMA Marion, 1870

Key: Wieser, 1953
Like *Pseudocella*, *Thoracostoma* has a strongly developed cephalic capsule with a hollow ventral tropis. However, true ocelli are present and the gubernacular apophysis is almost parallel to the spicules. The only British species, *T. coronatum*, is of common occurrence and is characterised by the cuticular granules on the posterior edge of the cephalic capsule (most other *Thoracostoma* species lack these structures) and the details of the male copulatory structures. However, the spicules and gubernaculum are sometimes difficult to make out easily because of the thickness and amount of musculature in this region of the body.

Species : 24

Thoracostoma coronatum (Eberth, 1863)
(Fig. 84)

Description. Body length 7.5–10.5 mm. Maximum diameter 150–195 μm (a = 38–65). Six labial papillae supported by conspicuous cuticularised subcuticular structures. Ten stout cephalic setae, 4–6 μm long. Six files of stout cervical setae with conspicuous innervation. Scattered short somatic setae present throughout the body. Head bears a conspicuous cephalic capsule, the posterior suture lying posterior to the amphid and lined with cuticular granules. The ventral section of the cephalic capsule resembles a hollow tooth, the tropis (Fig. 84A). Amphids 8 μm wide (0.16 c.d.). Buccal cavity narrow, conical with a dorsal tooth situated about level with the cephalic setae. The dorsal wall of the buccal cavity is well cuticularised and in optical section resembles a large hollow dorsal tooth. Ocelli situated about 110–120 μm from the anterior. Nerve ring situated 0.33 times oesophagus length from anterior. Tail short and rounded; caudal glands extend anterior to cloaca.

Spicules 115–130 μm with ventral alae.

Gubernaculum 90–100 μm long lying almost parallel to distal part of spicules but carrying a distinct solid knob-shaped apophysis. Ventral supplement situated about 100 μm anterior to cloaca. Two rows of 13–16 subventral papillae, each on a rounded protuberance; two situated posterior to cloaca. Testes paired, opposed.

Two opposed, reflexed ovaries. Vulva 63% of body length from anterior.

Distribution. West Ireland (under stones, amongst weed, on shore to 50 m); Plymouth (amongst weed and hydroids); Falmouth (in sponge and weed); Northumberland (kelp holdfasts); Isles of Scilly (amongst weed, sponges, hydroids, debris and kelp holdfasts).

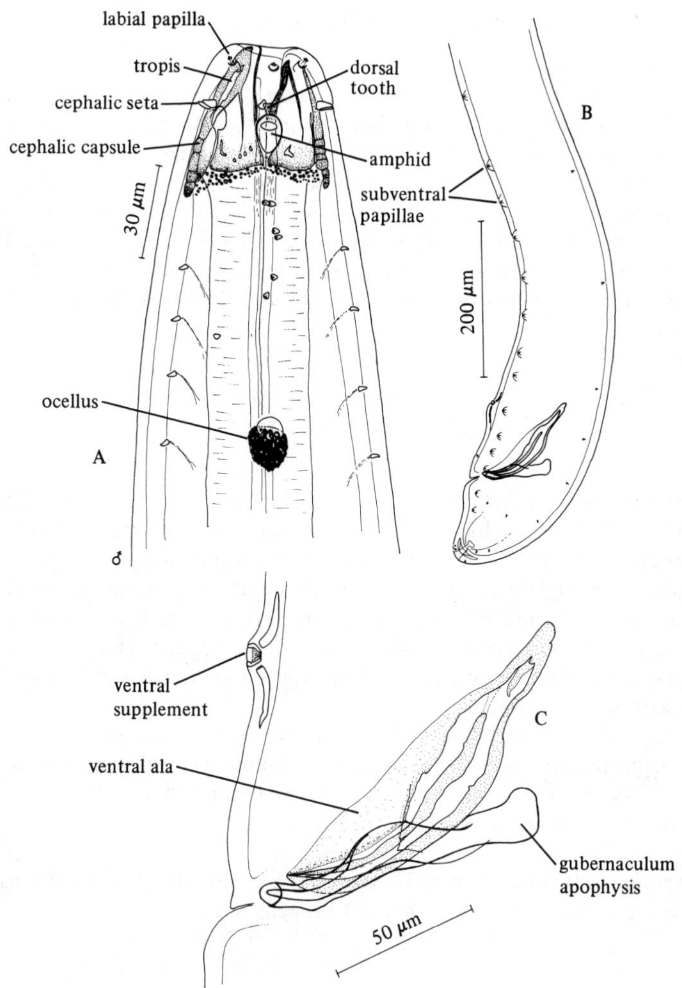

Fig. 84. *Thoracostoma coronatum*. A, Male head; B, Male tail; C, Copulatory apparatus. Original.

Genus HALALAIMUS De Man, 1888

Keys: Wieser, 1953; Mawson, 1958
The characteristic feature of *Halalaimus* is the long slit-like amphid and the strongly attenuated anterior end. The cuticle of some species can be seen to be faintly striated (perhaps all species have striated cuticles but some are beyond the resolving power of the light microscope). The striations depicted laterally on the anterior part of the tail in *H. isaitshikovi* (Fig. 85G) have also been reported from other species (e.g. *H. gracilis* by De Man, 1888 and *H. comatus* by Mawson, 1958). Reports of two other species require substantiation (*H. leptosoma* from Ireland and *H. longicollis* from North Sea kelp holdfasts).

Species : 61

Halalaimus gracilis De Man, 1888
(Fig. 85A–E)

Description. Body length 0.8–1.3 mm. Maximum diameter 17–26 μm (a = 40–60). Cuticle smooth, body setae absent. Cervical region long and attenuated. Six anterior cephalic setae 0.8–1.0 h.d. long; four posterior cephalic setae slightly longer and situated about 2 h.d. from the anterior. Buccal cavity absent. Amphids long and narrow; length about equal to distance from anterior end of body. Oesophagus length about 0.3 times body length with small posterior dilation. Tail long and slender, about 12–15 a.b.d. long, with slightly swollen tip.

Spicules 17 μm long (about 1.5 a.b.d.) with faint ventral ala.

Gubernaculum surrounds tip of spicules. Minute ventral spine anterior to cloaca. Two opposed testes (anterior left, posterior right of gut).

Two opposed ovaries. Vulva in middle of body.

Distribution. Plymouth (on red seaweed); Exe estuary (fine sand); Blyth estuary (intertidal mud). A common European species.

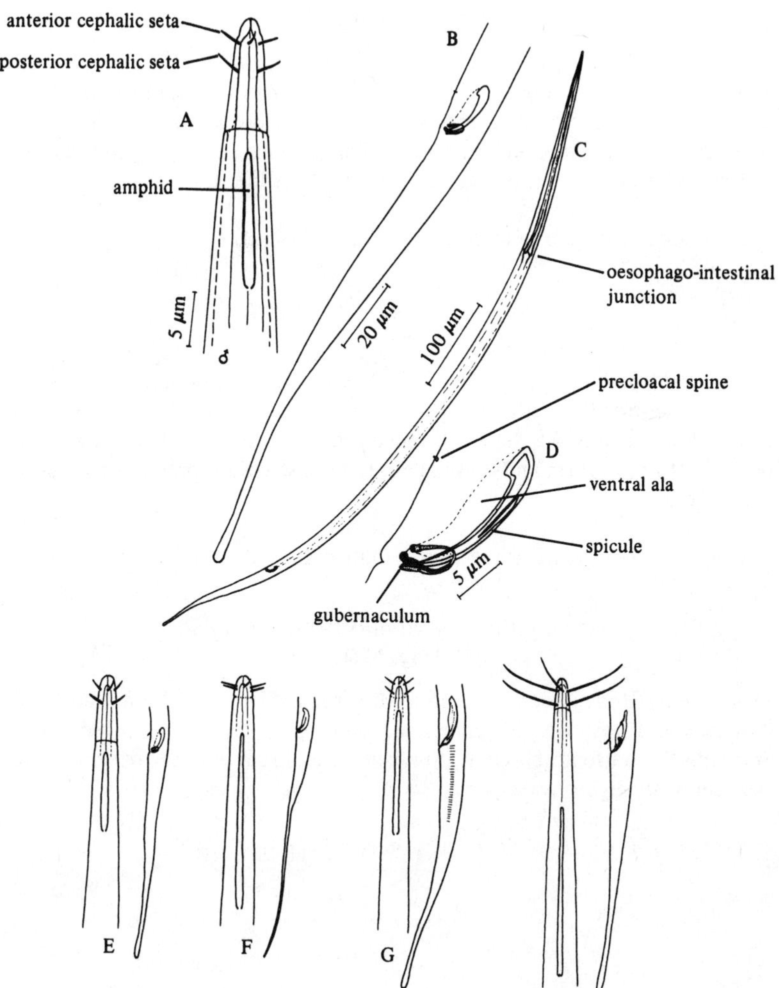

Fig. 85. *Halalaimus gracilis*. A, Male head; B, Male tail; C, Entire male; D, Spicules and gubernaculum. Original. E–H, Diagrammatic representations of head and male tail of E, *H. gracilis*, F, *H. longicaudatus*, G, *H. isaitshikovi*, H, *H. capitulatus*.

Halalaimus longicaudatus (Filipjev, 1927)
(Fig. 85F)

Description. Body length 2.0 mm. Maximum diameter 26 μm (a = 75). Four posterior cephalic setae shorter than but situated close to the six 0.9 h.d. long anterior cephalic setae. Amphids 48 μm long, about 3.5 times distance from anterior end of body. Tail long and tapering to a fine tip.

Distribution. Northumberland (fine sublittoral sand).

Halalaimus isaitshikovi (Filipjev, 1927)
(Fig. 85G)

Description. Body length 1.2 mm. Maximum diameter 28 μm (a = 44). Head distinctly set-off. Both circles of cephalic setae about equal length. Amphids 37 μm long, about 4.5 times distance from anterior of body. Tail with posterior third filiform and tip swollen. Lateral striations present on anterior part of tail.

Distribution. Northumberland (fine sublittoral sand).

Halalaimus capitulatus Boucher, 1977
(Fig. 85H)

Description. Body length 2.0 mm. Maximum diameter 12 μm (a = 165). Anterior and posterior cephalic setae both elongated, about 4 h.d. long. Amphids 43 μm long, about 1.3 times distance from anterior of body. Tail tapering with slightly swollen tip.

Distribution. Loch Ewe, Scotland (intertidal sand).

Genus OXYSTOMINA Filipjev, 1921

Key: Wieser, 1953
Oxystomina species have a characteristic oval-shaped amphid, with a curved structure round the anterior margin extending backwards as two projections. There is no buccal cavity. Typically there are six cephalic setae and four cervical setae, although these may be reduced and scarcely discernible in some species. Prominent oval cells are usually present, scattered throughout the body cavity. The excretory pore is usually conspicuous and strengthened by a cuticular thickening. The tail is clavate. Species are separated mainly on the size and position of the amphids and the cephalic and cervical setae.

In addition to the two species described below, there is an unsubstantiated record of *O. unguiculata* Stekhoven, 1935 from the Blyth estuary. This species is similar to *O. asetosa* except that the amphid is smaller and more anterior in position.

Species : 32

Oxystomina elongata (Bütschli, 1874)
(Fig. 86)

Description. Body length 1.8–2.0 mm. Maximum diameter 35–38 μm (a = 47–56). Cuticle smooth. Six cephalic setae 2.5 μm. Four cervical setae the same length, 2.75 h.d. from anterior. No buccal cavity. Amphids typical of genus, 3.5 μm wide, 29–30 μm from anterior in male, smaller and more rounded in female. Oesophagus 0.24 times body length, slender, with small posterior bulb. Ventral gland small, oval; excretory pore one-third of way down oesophagus length, surrounded by a cuticular ring. Tail 5 a.b.d. long, clavate.

Spicules 36–38 μm long (arc), with a slight bump ventrally just before the proximal tip.

Gubernaculum small and curved, enclosing spicule tips. One long and one short seta in front of cloaca.

Distribution. Blyth estuary; Skippers Island, Essex; Exe estuary (intertidal mud and muddy sand).

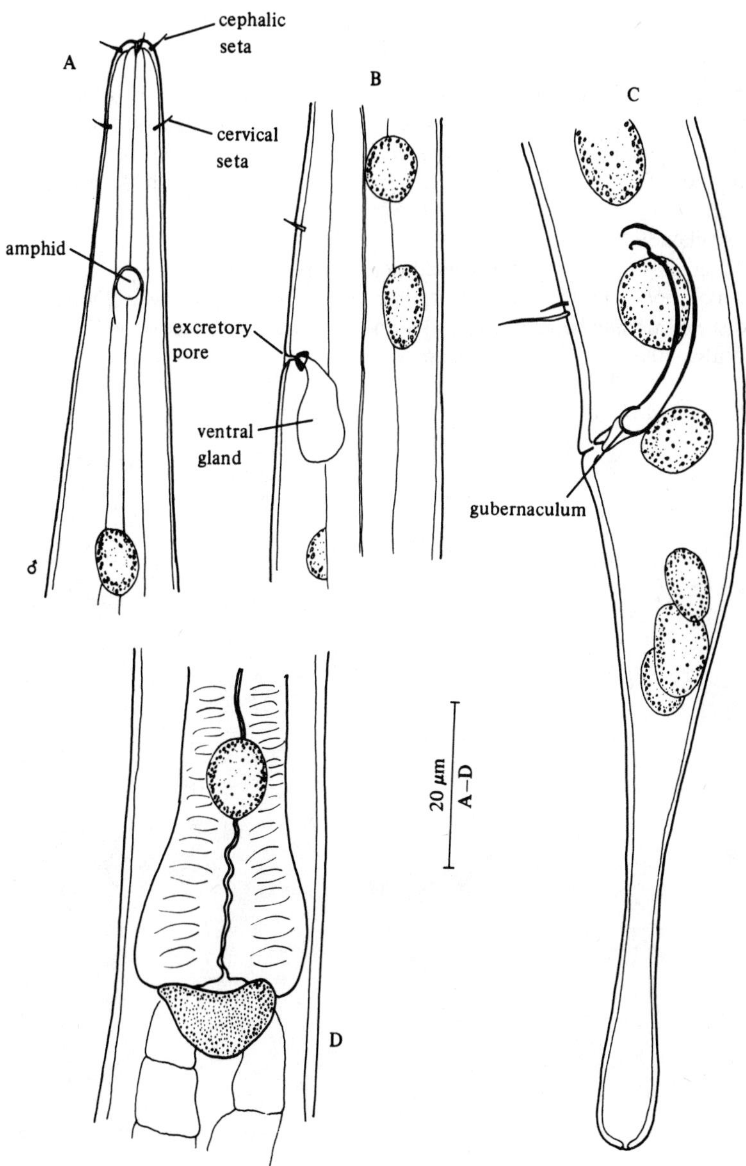

Fig. 86. *Oxystomina elongata*. A, Male head; B, Region of excretory pore; C, Male tail; D, Oesophageal bulb. Original.

Oxystomina asetosa (Southern, 1914)
(Fig. 87)

Description. Body length 3.4–4.0 mm. Maximum diameter 34–47 μm (a = 95–100). Very slender, tapers markedly at both ends. Cuticle smooth with no setae. Cephalic setae and cervical setae scarcely visible (or absent?). Amphids 15 μm long in both sexes. Excretory pore one-quarter of way down oesophagus length. Oesophagus 0.16 times body length. Tail 3.7–3.9 a.b.d., long, clavate.

Spicules 42 μm long, constricted near the proximal end.

Gubernaculum a small median plate with large rounded lateral lobes. An indistinct median precloacal papilla but no large precloacal setae.

Vulva 30% of body length from anterior.

Distribution. Clew Bay, West Ireland (sublittoral sand and shells).

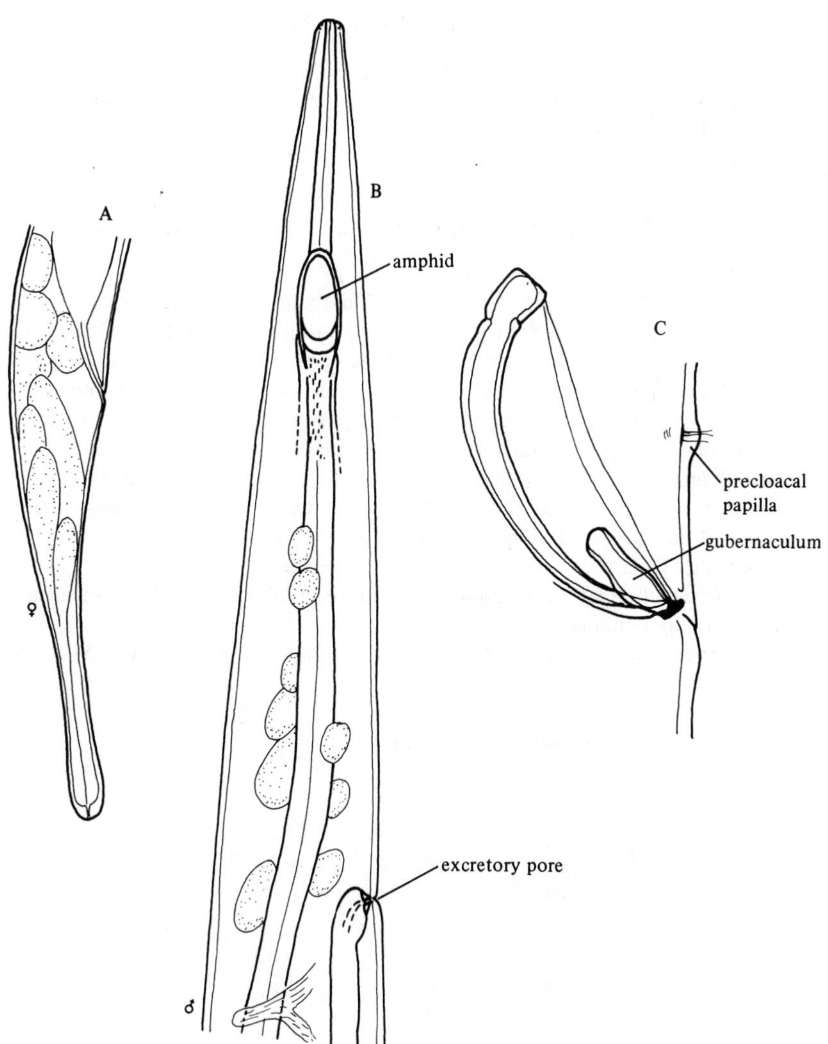

Fig. 87. *Oxystomina asetosa*. A, Female tail; B, Anterior end of male; C, Spicules and gubernaculum (from Southern, 1914).

Genus NEMANEMA Cobb, 1920

Key: Wieser, 1953 (as *Oxystomatina* subg. *Nemanema*)
Nemanema is almost identical to *Oxystomina* except that the tail is short and rounded, not clavate.

Species : 7

Nemanema cylindraticaudatum (De Man, 1922)
(Fig. 88)

Description. Body length 2.5–2.9 mm. Maximum breadth 32–56 μm (a = 52–78). Cuticle smooth. Six cephalic setae 2.5 μm, four cervical setae slightly shorter, 1.6 h.d. from anterior. No buccal cavity. Amphids as in *Oxystomina*, 2–2.8 times h.d. from anterior, 0.3 times c.d. wide in male, 0.24 times in female. Oesophagus slender, with small posterior bulb, 0.18 times body length. Excretory pore surrounded by thickened cuticular ring, 0.52 times oesophagus length from anterior in male, 0.3 times in female. Ventral gland elongate, extending to near base of oesophagus. Tail cylindrical with rounded tip, 2.6 a.b.d. long. Pseudocoel full of oval glandular cells.

Spicules 1.4 a.b.d. long, broad with rounded proximal ends.

Gubernaculum with a ring-like appearance in lateral view. No precloacal setae or supplements.

Anterior ovary vestigeal; posterior one large, reflexed. Vulva at 34% of body length.

Distribution. Exe estuary (intertidal mud).

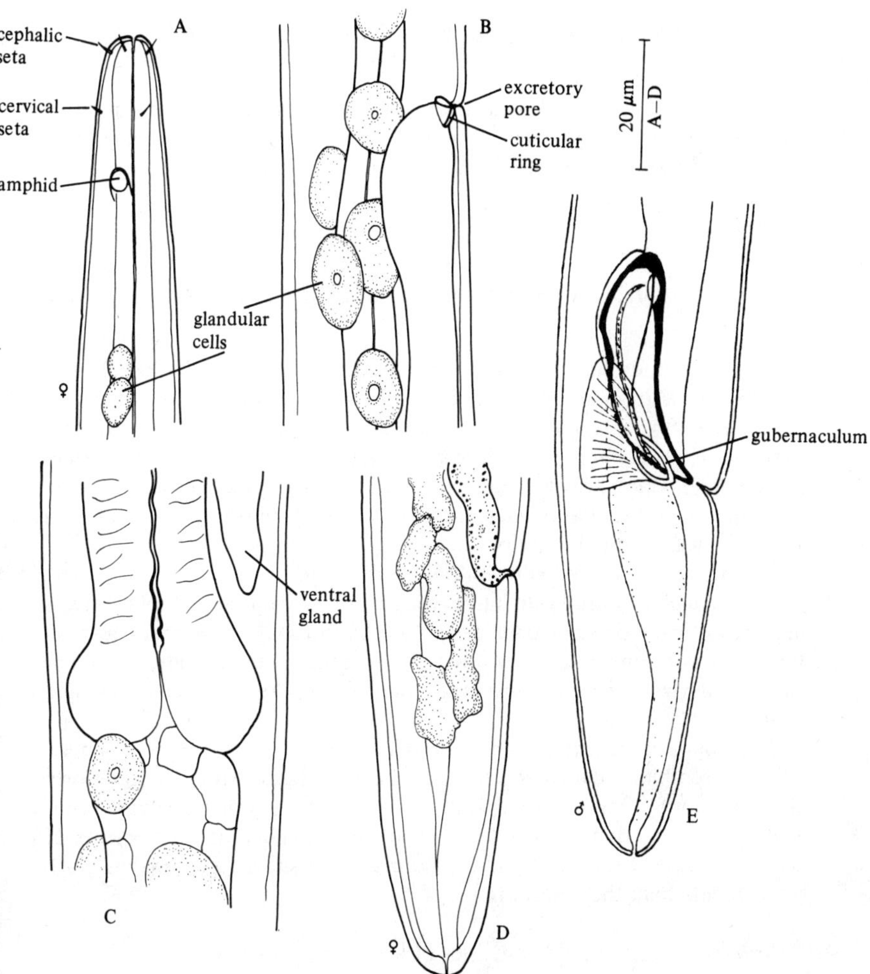

Fig. 88. *Nemanema cylindraticaudatum*. A, Female head; B, Region of excretory pore; C, Oesophageal bulb; D, Female tail (original); E, Male tail (from Stekhoven, 1935).

Genus PAROXYSTOMINA Micoletzky, 1924

Key: Wieser, 1953
Males of *Paroxystomina* are distinctive in having two subventral rows of winged precloacal supplements. Apart from *P. asymmetrica*, the only other species so far known is *P. micoletzkyi* Wieser, 1953 from littoral algae off the coast of Chile; it has 4 precloacal supplements on each side.

Species : 2

Paroxystomina asymmetrica Micoletzky, 1924
(Fig. 89)

Description. Body length 2.3–4.7 mm. Maximum diameter 40–90 μm (a = 52–118). Cuticle smooth, body setae absent. Cervical region long and attenuated. Six anterior cephalic setae about 1 μm long; four posterior cephalic setae 2–3 μm long. Rows of lateral and medial cervical setae, 3–4 μm long. Buccal cavity small, conical. Amphids situated posterior to the four cephalic setae. Oesophagus narrow with slight posterior dilation. Tail short, conical with pointed tip.

Spicules about 40 μm with ventrally curved tip.

Gubernaculum paired with lateral wings. Two subventral pairs of precloacal spines situated on small papillae and one pair of subventral spines near tail tip. Two subventral rows of precloacal winged supplements; 4 in left subventral row, 10 in right subventral row. Two opposed testes, both situated to the right of the gut.

Two opposed, reflexed ovaries. Prominent vulva has two rows of supplements resembling cuticularised rings bearing pegs, situated anteriorly and posteriorly. One row follows the ventral line, the other is situated subventrally to the right of the body. The total number of supplements in each row varies between 3 and 11; normally the subventral row has more supplements than the ventral row.

Distribution. Northumberland; Isles of Scilly (kelp holdfasts).

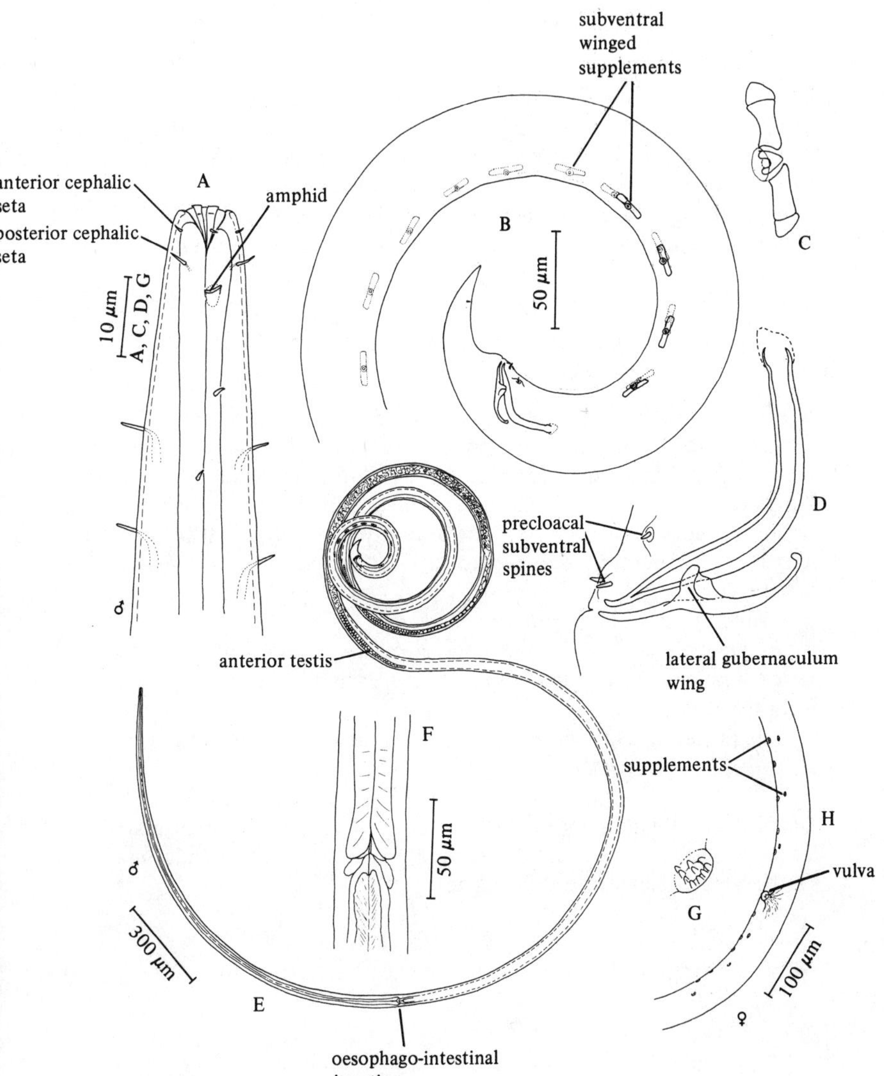

Fig. 89. *Paroxystomina asymmetrica*. A, Male head; B, Posterior end of male; C, Precloacal supplement; D, Spicules and gubernaculum; E, Entire male; F, Posterior oesophageal region; G, Vulval supplement; H, Vulva region. Original.

Genus THALASSOALAIMUS De Man, 1893

Key: Wieser, 1953
Thalassoalaimus species have large pocket-like amphids, ten or twelve cephalic setae and four cervical setae. There is no buccal cavity. Only one species, *T. tardus*, has been recorded from the British Isles, but others are likely to occur. Species vary in tail shape and some have a thick cuticular lining to the tail tip, called the caudal capsule (absent in *T. tardus*). They also vary in the length and position of the cephalic and cervical setae.

Species : 21

Thalassoalaimus tardus De Man, 1893
(Fig. 90)

Description. Body length 1.5–3.2 mm (a = 55–102). Cuticle smooth. Twelve cephalic setae arranged in two circles, all 0.5–0.8 h.d. long. Four cervical setae slightly longer than cephalic setae, 3.6–5.3 times h.d. from anterior. Amphids pocket-like with elliptical openings, immediately behind cephalic setae, 0.5 times c.d. wide. No buccal cavity. Oesophagus 0.14 times body length, with small posterior bulb. Excretory pore not prominent. Tail conical, 2.2–3.0 a.b.d. long.

Spicules 1 a.b.d. long, broad.

Gubernaculum about half as long as spicules. Male with two precloacal raised supplements, about as far apart from each other as the posterior one is from the cloaca.

Only posterior ovary present, reflexed, with a small pre-vulvar sac. Vulva at 24–30% of body length.

Distribution. Plymouth (intertidal algae); Blyth estuary (intertidal mud); Skippers Island, Essex (intertidal mud); Exe estuary (intertidal mud and sand); Isles of Scilly (intertidal algae).

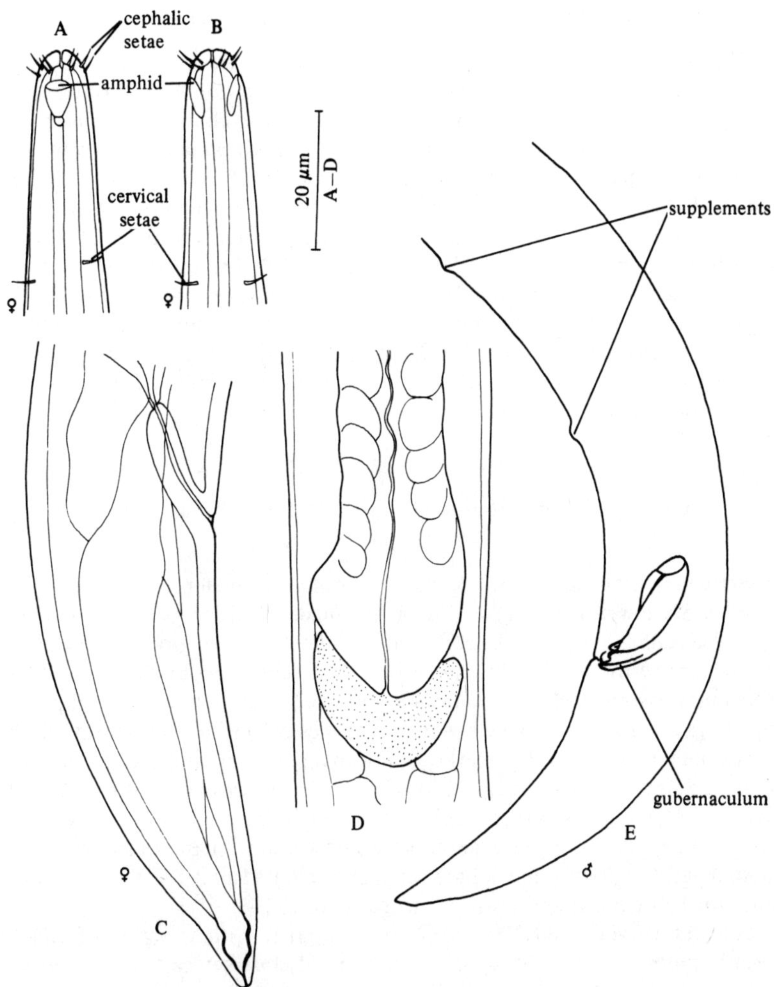

Fig. 90. *Thalassoalaimus tardus*. A, Female head (lateral); B, Female head (dorsal); C, Female tail; D, Oesophageal bulb (original); E, Posterior end of male (from Stekhoven, 1935).

Genus ADONCHOLAIMUS Filipjev, 1918

Adoncholaimus has been distinguished from *Viscosia* in the past by the presence of a gubernaculum in the male. This is a doubtful character, as muscular structures associated with the anus have almost certainly been mistaken for a gubernaculum in some species. *Adoncholaimus*, however, has long slender spicules in comparison with *Viscosia*, and a much more complex and well developed demanian system in the female, with paired or multiple exit pores positioned laterally a short distance preanal, which are absent in *Viscosia*. *Adoncholaimus lepidus* (De Man, 1889) has not been recorded from Britain but is common in brackish water in the Baltic and on the Dutch and German coasts and may well turn up. It is superficially similar to *A. panicus* but smaller (2.4–2.9 mm), with relatively shorter spicules (2.6 a.b.d.) and the demanian system opens by *single* lateral pores, a little over one anal diameter preanal.

Species: 16

Adoncholaimus fuscus (Bastian, 1865)
(Fig. 91)

Description. Body length 3.7–4.9 mm. Maximum diameter 103–142 μm (a = 33–37). Six low rounded lips with small rounded labial papillae. Ten small rounded cephalic papillae. Buccal cavity divided by a deep groove at the level of the tips of the smaller buccal teeth. Right subventral tooth longer than remaining two, which are equal and prominent. Amphids with elliptical openings and rounded pockets, 8 μm (0.2 times c.d.) wide. Oesophagus 0.16 times body length, swollen markedly towards posterior end, but no definite bulb. Excretory pore about 1.15 times the buccal cavity length from the anterior. Nerve ring 0.46 times oesophagus length from anterior. A few short rounded papillae over the general body surface, most numerous in anterior oesophageal region. A few longer setae on the tail. Tail 2.1–2.3 a.b.d. long, anterior half conical and posterior half cylindrical.

Spicules curved, 180–190 μm (2.3–2.5 a.b.d.), with a series of small dorsal spines near the distal tip. Four pairs of close-set papillae on either side of cloaca, and two files of 7–8 papillae more lateral in position extending to base of conical part of tail.

Gubernaculum flat, plate-like, 39–45 μm long.

Ovaries paired, equal, reflexed. Single lateral exit pores of demanian system 1.8 a.b.d. preanal, heavily cuticularised and bordered by a fan of cells (moniliform glands). Vulva at 51–52% of body length.

Distribution. Found at several localities in England and Scotland. Typical of muddy or sandy intertidal sediments in areas of reduced salinity.

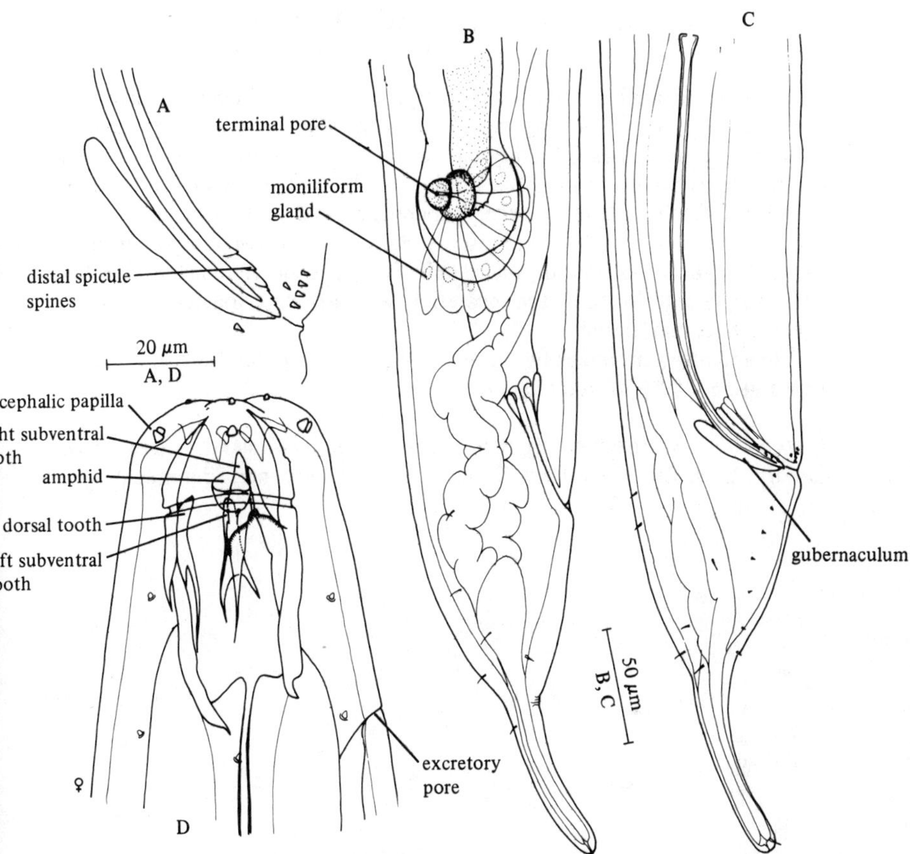

Fig. 91. *Adoncholaimus fuscus*. A, Cloacal region of male; B, Female tail; C, Male tail; D, Female head. Original.

Adoncholaimus thalassophygas (De Man, 1876)
(Fig. 92)

Description. A smaller species; body length 1.7–2.2 mm. Maximum diameter 48–68 μm (a = 31–37). Arrangement of cephalic papillae and buccal teeth as in *A. fuscus*. Amphids 0.3 times c.d. wide. Excretory pore 1.4–1.8 buccal cavity lengths from anterior. Male tail conical for its proximal two-fifths, then cylindrical; female tail more evenly tapered, 3 a.b.d. in both sexes.

Spicules slender, 82–94 μm (3 a.b.d.). Paired rounded ventral swellings at base of conical portion of male tail. Six pairs of circumcloacal setae subventrally anterior to cloaca, posterior two pairs close together.

Gubernaculum absent.

Demanian system opens by a pair of single lateral pores 4 a.b.d. preanal. Vulva at 56 – 59% of body length.

Distribution. Whitstable; Blythe estuary; Skippers Island, Essex; Exe estuary; Loch Etive, Scotland. Typical of low salinity intertidal mud.

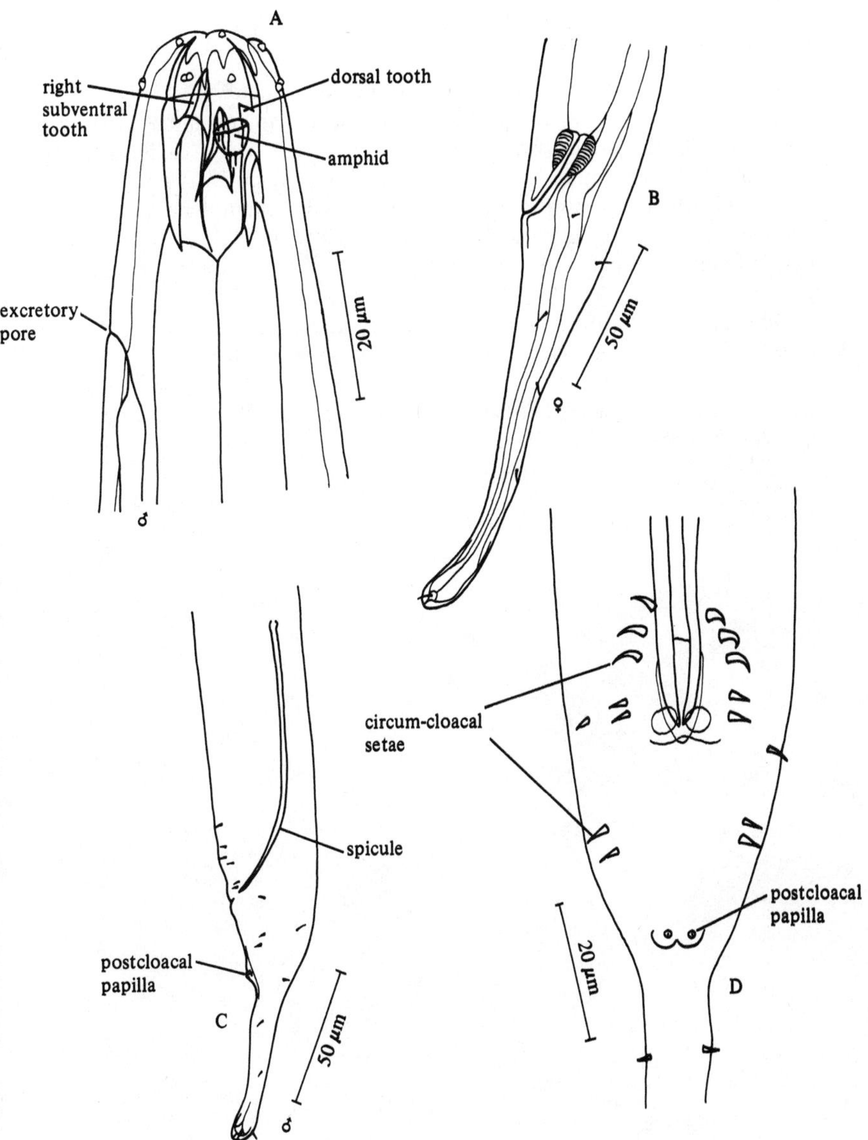

Fig. 92. *Adoncholaimus thalassophygas*. A, Male head; B, Female tail; C, Male tail; D, Ventral view of cloacal region. Original.

Adoncholaimus panicus Cobb, 1930
(Fig. 93)

Description. Body length 40–46 mm. Maximum diameter 93–135 μm (a = 43–59). Head as in other species but cephalic papillae more pointed. Amphids 0.18 times c.d. wide. Excretory pore 1.5–2 buccal cavity lengths from anterior. Tail similar in shape to *A. thalassophygas*, 3.1–4 a.b.d. long.

A small structure posterior to the spicule tips is present in some specimens, and may represent the gubernaculum. Spicules 200 μm (3.3–4 a.b.d.). Four pairs of small papillae anterior to cloaca, more lateral papillae on conical portion of tail.

Demanian system opens on either side by a transverse row of usually 8 pores 1.75–2 a.b.d. preanal. Vulva at 50% of body length.

Distribution. Loch Etive, Scotland (muddy intertidal). Wrongly designated by Warwick and Gage (1975) as *A. lepidus* (De Man, 1889).

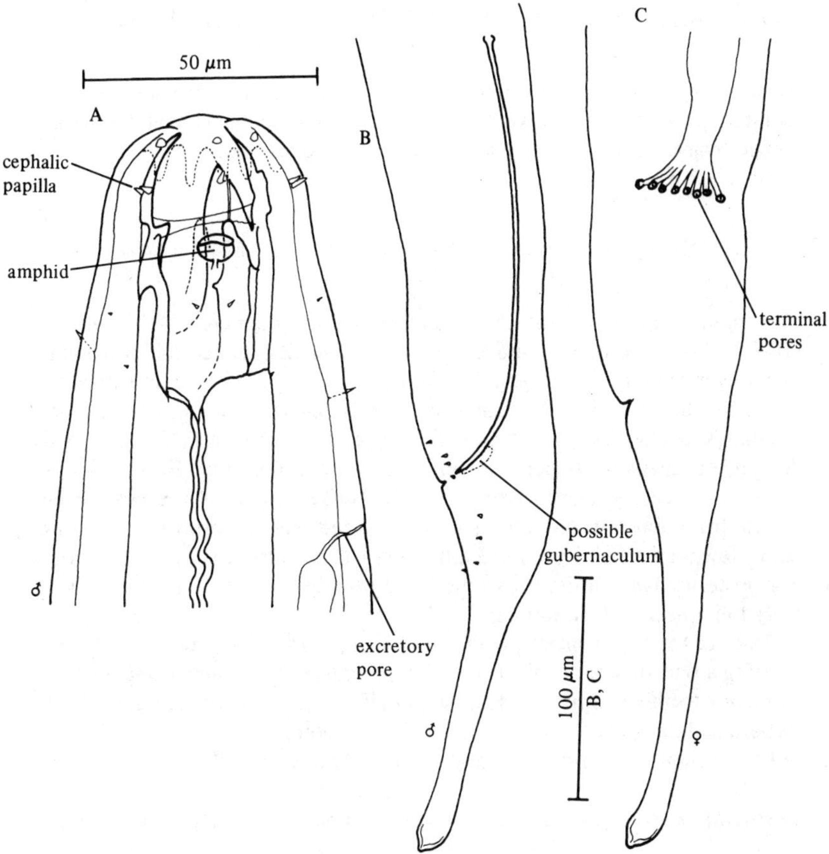

Fig. 93. *Adoncholaimus panicus*. A, Male head; B, Male tail; C, Female tail. Original.

Genus METAPARONCHOLAIMUS De Coninck and Stekhoven, 1933

Metaparoncholaimus is distinguished from other oncholaimid genera by the combination of equal subventral teeth (longer than the dorsal tooth) and single ovary. *M. campylocercus* is the only European species.

Species : 6

Metaparoncholaimus campylocercus (De Man, 1876)
(Fig. 94)

Description. Body length 2.5–5.3 mm. Maximum diameter 44–88 μm (a = 58–78). Six prominent rounded lips with six small rounded labial papillae. Six longer cephalic setae 4 μm (0.17 h.d.), shorter four 3 μm. Two subventral teeth in buccal cavity of equal size, slightly bigger than the dorsal tooth. Amphids pocket-like, 0.3 times c.d. wide. Oesophagus 0.15 times body length, broadens at its posterior end, but no definite bulb. Excretory pore 2.75 buccal cavity lengths from anterior. Nerve ring 0.5 times oesophagus length from anterior. Numerous short scattered setae anterior to nerve ring, and a few on the tail. Male tail 3 a.b.d. long, conical for first third, cylindrical for posterior two-thirds; tip swollen and rounded. Female tail shorter with only the apical half cylindrical.

Male tail with a ventral swelling three-fifths of the way down its length, bearing a pair of stout conical subventral papillae and a pair of smaller setae posterior to them. Spicules with distal half swollen, 32 μm (1.5 a.b.d.). No gubernaculum. Seven pairs of circumcloacal setae.

Ovary single, anterior. Vulva at 58–61% of body length.

Distribution. A single male from intertidal sand at low water, Exe estuary.

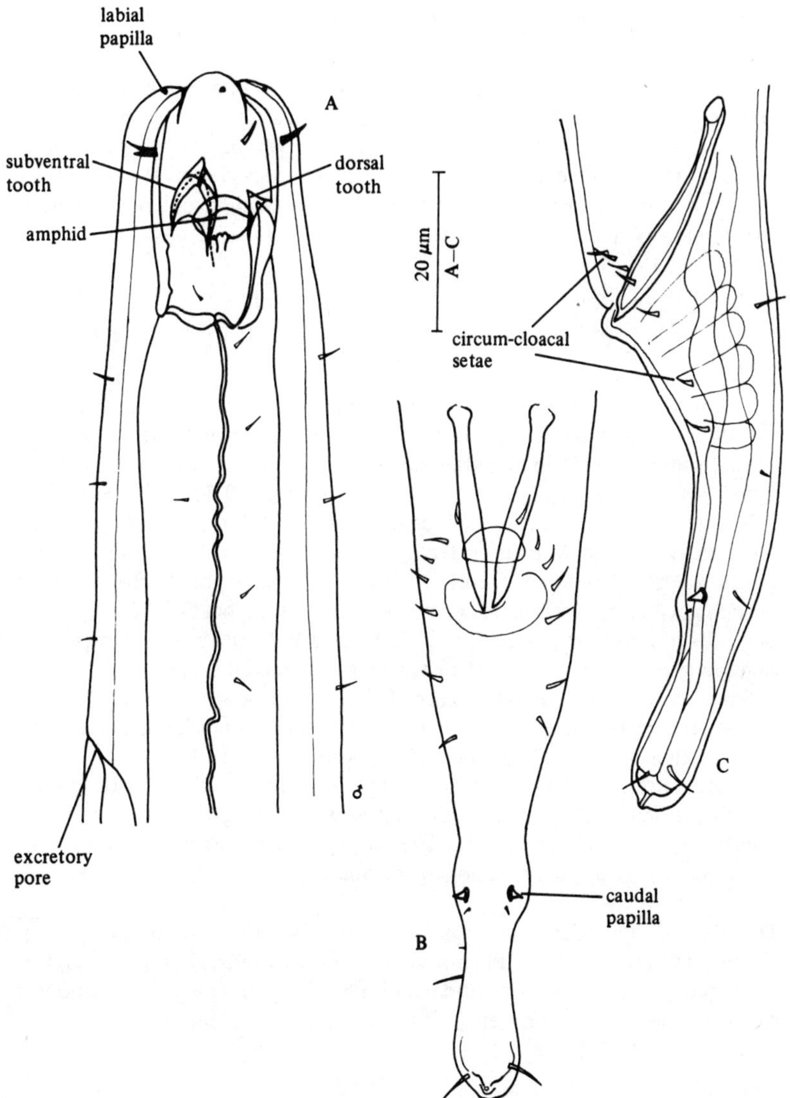

Fig. 94. *Metaparoncholaimus campylocercus*. A, Male head; B, Male tail (ventral); C, Male tail (lateral). Original.

Genus METONCHOLAIMUS Filipjev, 1918

Key: Chitwood, 1960
Metoncholaimus is similar to *Oncholaimus* except that in most cases the spicules are elongated and slender (e.g. *M. albidus*). A few species, however, do have short spicules (e.g. *M. scanicus*), but in this case a gubernaculum is present, which is absent in *Oncholaimus.* The two species known so far from British waters can easily be distinguished by the spicule length of the male or the demanian system of the female.

Species : 15

Metoncholaimus albidus (Bastian, 1865)
(Fig. 95)

Description. Body length 5.4–6.7 mm. Maximum diameter 78–110 μm (a = 54–75). Cuticle smooth. Cervical setae present anterior to nerve ring and caudal spines present in male but somatic setae otherwise absent. Six small labial setae. Ten cephalic setae; six 9.5–11 μm (0.26–0.30 h.d.); four 7–9.5 μm (0.20–0.25 h.d.). Amphids level with dorsal tooth, 11–13 μm wide. Buccal cavity large with three teeth, the left subventral being the largest. Excretory pore and nerve ring at about 0.14–0.16 and 0.47–0.49 times oesophagus length from anterior respectively. Oesophagus 0.10–0.12 times body length, widens posteriorly. Tail 5–6.5 a.b.d.; male with conspicuous oblique musculature causing the characteristic hooked shape.

Spicules elongated, about 400 μm (6.6–8.0 a.b.d.).

Gubernaculum present. Prominent cloacal bulge (not included in measurement of cloacal body diameter). Two opposed testes.

Single anterior ovary. Vulva at 64–68% of body length. Anterior to anus (in mature females) lies a cuticularised girdle carrying the lateral terminal pores of the demanian system. The uvette appears as a rosette of cells about 350 μm (7 a.b.d.) anterior to the anus.

Distribution. Falmouth (amongst small stones and sand in tide pools); Cornwall (intertidal sand); Plymouth (amongst hydroids); Essex coast (intertidal mud); Isles of Scilly (intertidal and sublittoral, in sediment and amongst weed, hydroids, bryozoans etc.); Plymouth (sublittoral mud).

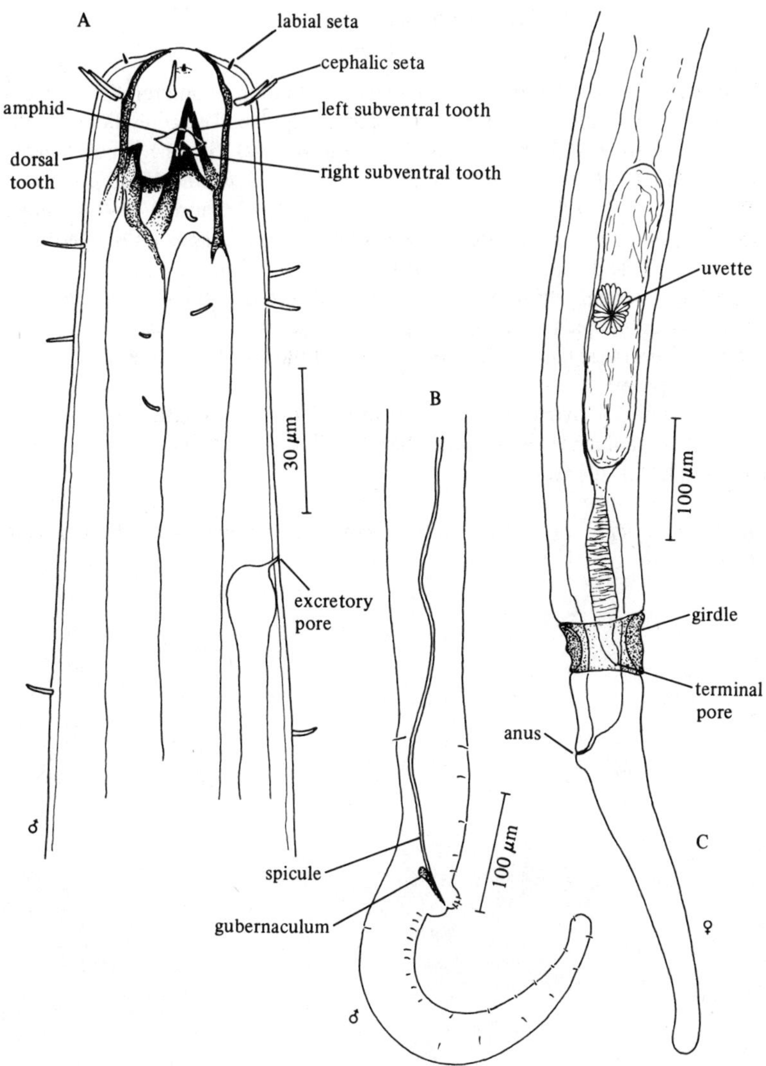

Fig. 95. *Metoncholaimus albidus*. A, Anterior end of male; B, Posterior end of male; C, Posterior end of female. Original.

Metoncholaimus scanicus (Allgén, 1935)
(Fig. 96)

Description. Body length 9.4–10.9 mm. Maximum diameter 110–118 μm (a = 85–93). Cuticle smooth. Six small labial papillae. Ten cephalic setae; six 12.5–15 μm (0.30–0.38 h.d.); four 11–13 μm (0.26–0.31 h.d.). Buccal cavity similar to *M. albidus*. Amphids level with dorsal tooth, 13–15 μm wide. Excretory pore and nerve ring at about 0.20–0.23 and 0.41–0.45 times oesophagus length from anterior respectively. Oesophagus 0.08–0.09 times body length. Tail same shape in both sexes, 3.2–3.5 a.b.d.

Spicules slender, straight, 93–110 μm (1.9 a.b.d.). Small gubernaculum present at tip of spicules. About 12 pairs of setae around cloaca. Two pairs of short subventral setae about two-thirds of tail length from cloaca.

Two opposed testes.

Single anterior ovary. Vulva at about 69% of body length. Paired lateral pores of demanian system situated about 5 a.b.d. from anus and uvette about 680 μm (12 a.b.d.) from anus.

Distribution. Isles of Scilly (subtidal sand and among hydroids, bryozoa, etc.).

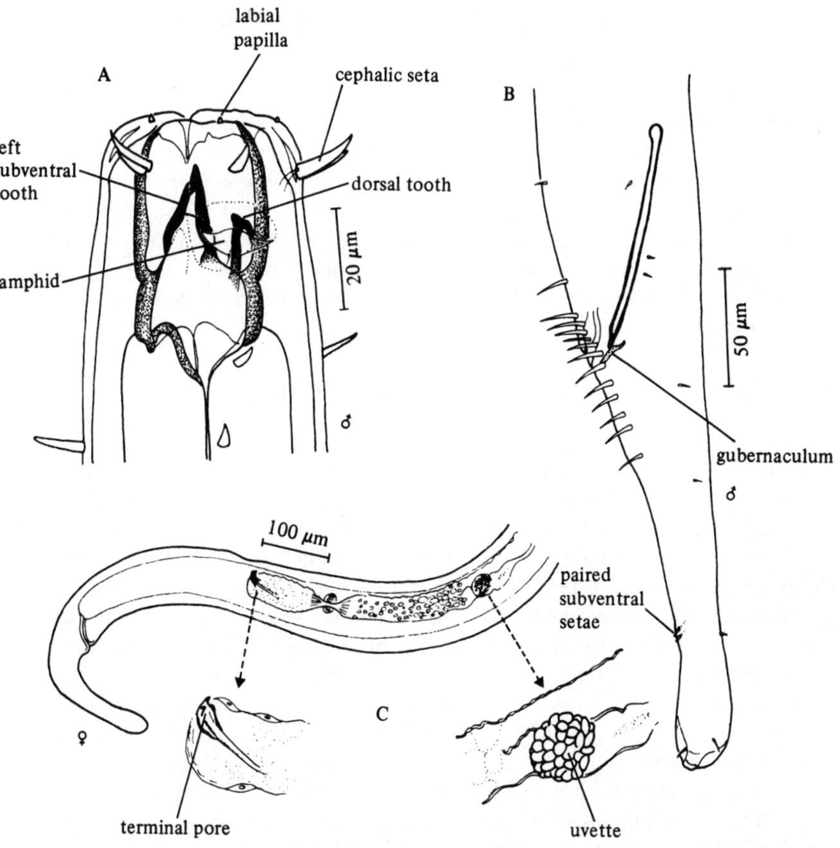

Fig. 96. *Metoncholaimus scanicus*. A, Male head; B, Male tail; C, Posterior end of female with details of demanian system. Original.

Genus ONCHOLAIMELLUS De Man, 1890

Keys: Wieser, 1953; Timm, 1969
The genus is characterised by a massive right subventral tooth and a transverse band dividing the buccal cavity into two halves. Only the two species described here have unequal spicules and three of the other species lack a bursa copulatrix. *O. calvadosicus* and *O. mediterraneus* can be distinguished mainly on the relative lengths of the cephalic setae, spicules and tail and on the structure of the bursa. *O. mediterraneus* has only been described previously from the Mediterranean, from which the specimens described here from Northern Ireland differ only in minor details.

Species: 7

Oncholaimellus calvadosicus De Man, 1890
(Fig. 97)

Description. Body length 1.9–2.1 mm. Maximum diameter 24–50 μm (a = 43–84). Cuticle smooth. Six low rounded lips each with a papilliform sensillum. Male with six 16–17 μm (1 h.d.) and four 11–12 μm (0.6 h.d.) cephalic setae. Female with all ten cephalic setae 7–9 μm (0.5 h.d.) long. A few short setae in anterior oesophageal region but otherwise somatic setae absent. Head constricted just behind cephalic setae. Amphids indistinct. Buccal cavity in two parts, separated by cuticularised band: three teeth in posterior part, the right subventral being the biggest. Oesophagus cylindrical. Excretory pore 0.5 times oesophagus length from anterior. Tail about 5.5 a.b.d., cylindrical with swollen tip. Caudal glands extend well anterior to anus.

Spicules unequal: right 80–100 μm (5.5 a.b.d.), left 45–50 μm (3 a.b.d.). Gubernaculum absent. Copulatory bursa present consisting of two longitudinal wings of cuticle occupying anterior third of tail. Paired setae present at both ends of bursa and a pair of supporting rods about one-third of its length from cloaca.

Vulva at 44–46% of body length. Two opposed, reflexed ovaries.

Distribution. Whitstable (intertidal sandflat); Exe estuary (intertidal sand). A common European species.

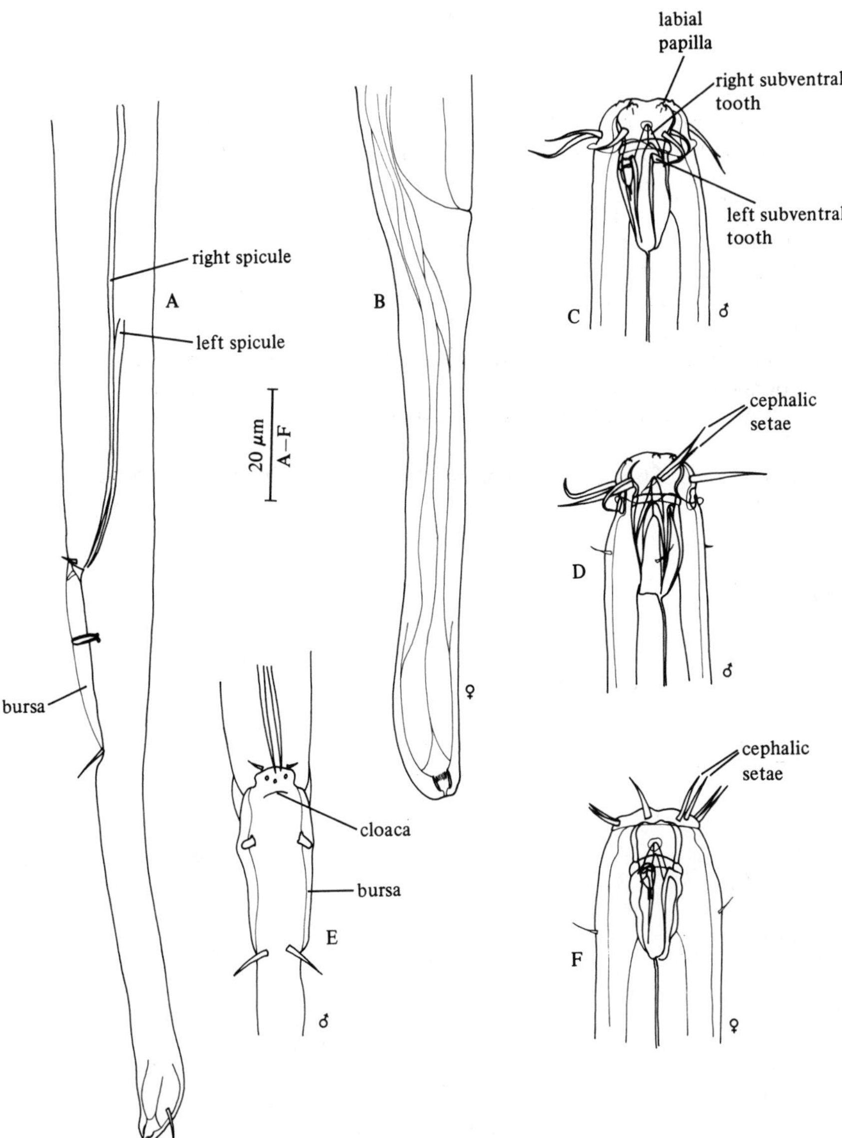

Fig. 97. *Oncholaimellus calvadosicus*. A, Male tail; B, Female tail; C, Male head (lateral); D, Male head (dorsal); E, Cloacal region (ventral); F, Female head (lateral). Original.

Oncholaimellus mediterraneus Stekhoven, 1942
(Fig. 98)

Description. Body length 1.3–1.5 mm. Maximum diameter 35–40 μm (a = 34–36). Cuticle smooth. Six lips each with a minute seta. Male with six 5–7 μm (0.5–0.6 h.d.) and four 5–6 μm (0.4 h.d.) cephalic setae. Female with six 4–5 μm (0.3 h.d.) and four 3–4 μm (0.25 h.d.) cephalic setae. Somatic setae absent. Head constricted behind cephalic setae. Amphids indistinct. Buccal cavity as in *O. calvadosicus.* Excretory pore 0.57–0.60 times oesophagus length from anterior. Tail 3–4 a.b.d., cylindrical with swollen tip.

Spicules unequal; right 74–77 μm (4.4–5.0 a.b.d.), left 53–54 μm (3–3.5 a.b.d.).

Gubernaculum absent. Copulatory bursa present, supported and surrounded by a complex arrangement of spines, setae and papillae.

Vulva at 48% of body length. Two opposed, reflexed ovaries.

Distribution. Strangford Lough, Northern Ireland (intertidal sandflat).

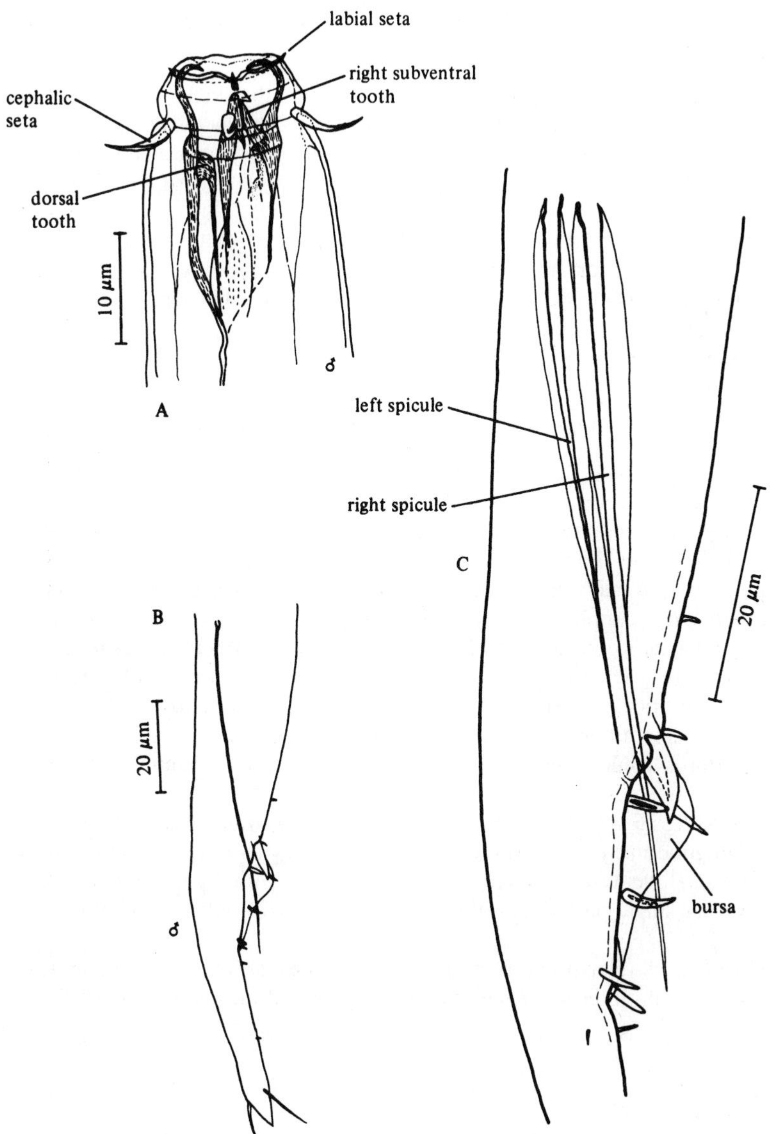

Fig. 98. *Oncholaimellus mediterraneus*. A, Male head; B, Male tail; C, Spicules and bursa copulatrix. Original.

Genus ONCHOLAIMUS Dujardin, 1845

Key: Wieser, 1953
The combination of large left ventrolateral tooth, single ovary and short spicules with no gubernaculum serves to distinguish *Oncholaimus* from other oncholaimid genera. Two groups of species are easily separated on the basis of tail shape: *O. brachycercus, O. dujardinii* and *O. oxyuris* with short (less than 2.5 a.b.d.) ventrally curved tails, and *O. campylocercoides*, *O. skawensis* and *O. attenuatus* with longer (3 or more a.b.d.) straighter tails. The accessory papillae and circumcloacal setae associated with copulation in the male are useful characters for distinguishing the species within these groups.

Species: 75

Oncholaimus brachycercus De Man, 1889
(Fig. 99)

Description. Body length 3.0–4.3 mm. Maximum diameter 40–65 μm (a = 57–82). Six rounded lips with six small conical labial setae. Six longer cephalic setae 7 μm (one-quarter of h.d.), remaining four only slightly shorter. Left subventral tooth in buccal cavity slightly longer than the other two. Amphids pocket-like, 8–10 μm wide (one-quarter of c.d.). Oesophagus cylindrical, 0.125 times body length. Excretory pore 2.2–3.2 buccal cavity lengths from anterior. Nerve ring 0.5 times oesophagus length from anterior. Short setae scattered over general body surface, most numerous in anterior region and on tail. Tail short, 1.2–1.9 a.b.d., anterior two-thirds conical, posterior third cylindrical, always ventrally curved in male.

Spicules 36–39 μm (a little over 1 a.b.d.), fairly straight, distal half dilated. No gubernaculum. Seven or eight pairs of long stout circumcloacal setae. Two small ventral bumps near the tail tip, each with a pair of short setae.

Ovary single, anterior, reflexed. Vulva at 72–73% of body length.

Distribution. Recorded from several localities around the British Isles. Usually from intertidal sand, but also among hydroids and seaweeds.

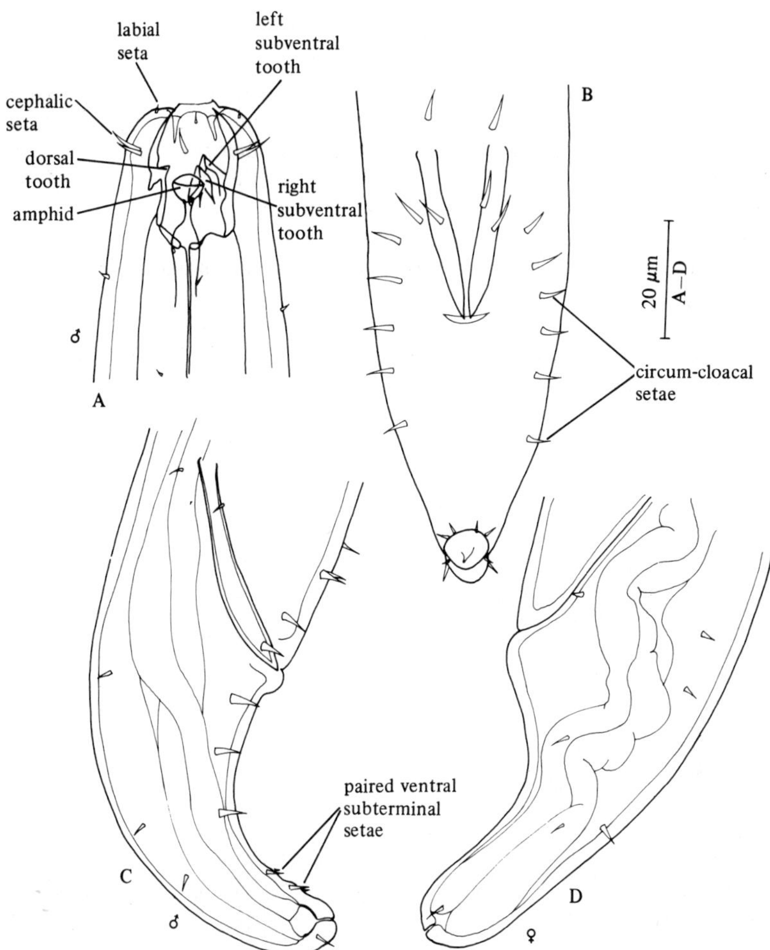

Fig. 99. *Oncholaimus brachycercus*. A, Male head; B, Male tail (ventral); C, Male tail (lateral); D, Female tail. Original.

Oncholaimus dujardinii De Man, 1876
(Fig. 100)

Description. Body length 2–2.2 mm. Maximum diameter 40–48 μm (a = 45–49). Six cephalic papillae small and rounded. Longer cephalic setae 0.14 times h.d. Left subventral tooth larger than other two. Excretory pore 2.75 buccal cavity lengths from anterior. Tail strongly attenuated behind the anus, 2.3 a.b.d. long, ventrally curved in both sexes.

Spicules swollen in distal half, 1.75 a.b.d.

No gubernaculum. Six or seven pairs of curved, thorn-like circumcloacal setae.

Distribution. North-east England; Plymouth; Isles of Scilly (intertidal seaweeds and holdfasts).

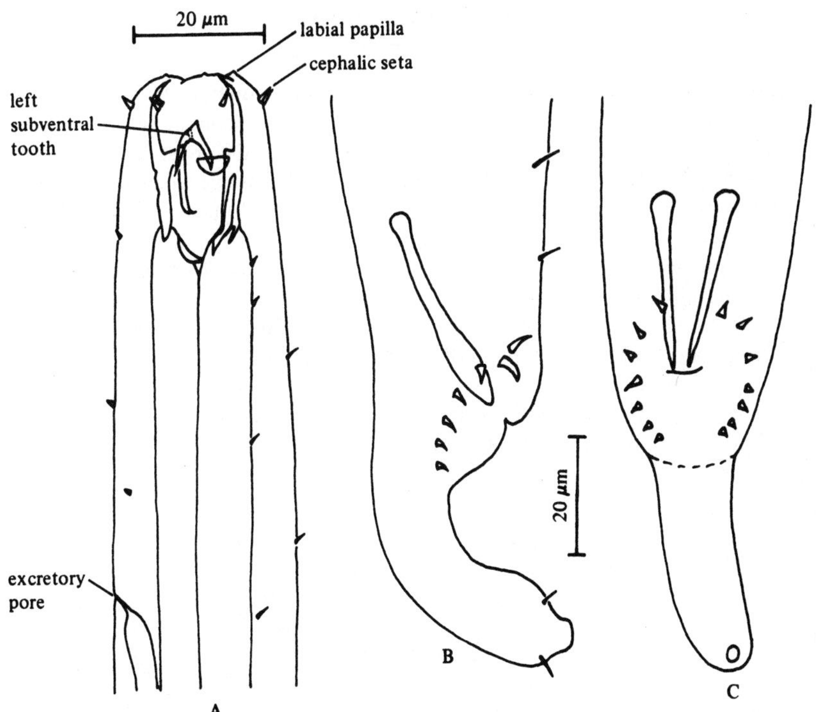

Fig. 100. *Oncholaimus dujardinii*. A, Male head (from Stekhoven, 1950); B, Male tail (lateral); C, Male tail (ventral) (from Inglis, 1962).

Oncholaimus oxyuris Ditlevsen, 1911
(Fig. 101)

Description. Body length 2.3–2.5 mm. Maximum diameter 50–59 μm (a = 42–46). Labial papillae small and rounded. Six longer cephalic setae 7–8 μm (0.25 h.d.), shorter four 5–6 μm. Left subventral tooth longer than others. Amphids 0.24 times c.d. wide. Excretory pore 1.1–2.6 buccal cavity lengths from anterior. Tail 1.3–1.6 a.b.d. long, anterior half conical, posterior half cylindrical.

A large ventral swelling about three-fifths of the way down the length of the male tail.

Spicules slightly bent, swollen in distal half, 47–50 μm (1.4 a.b.d.).

No gubernaculum. 8–10 pairs of circumcloacal setae.

Distribution. Barrow-in-Furness (power station intake); Exe estuary (coarse and muddy sand near high water mark).

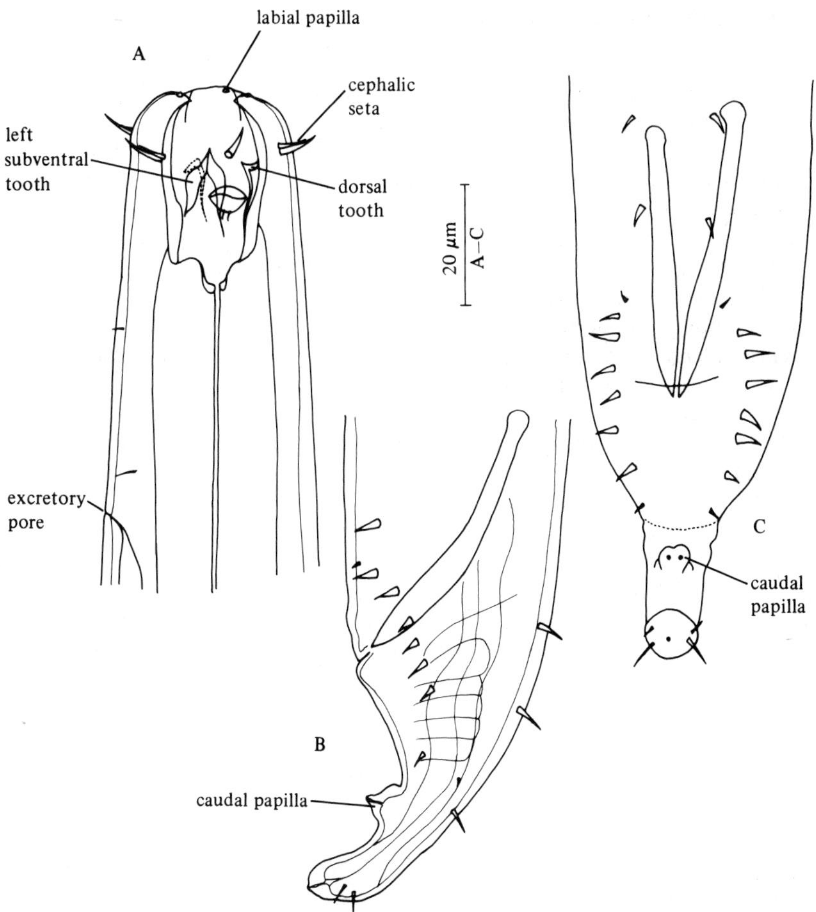

Fig. 101. *Oncholaimus oxyuris*. A, Male head; B, Male tail (lateral); C, Male tail (ventral). Original.

Oncholaimus campylocercoides De Coninck and Stekhoven, 1933
(Fig. 102)

Description. Body length 3.6–4.3 mm. Maximum diameter 58–80 μm (a = 47–62). Six small conical labial setae. Six longer cephalic setae 8 μm (0.23 h.d.), shorter four 6–7 μm. Left subventral tooth larger than other two. Amphids 11–12 μm wide (one-third of c.d.) with reniform openings and shallow pockets. Excretory pore 2.0–2.8 buccal cavity lengths from anterior. Tail 2.2–3.0 a.b.d., anterior third to half conical, posterior part cylindrical.

Spicules 46–48 μm (1.3 a.b.d.), straight, distal half swollen.

No gubernaculum. Nine or ten pairs of circumcloacal setae. Two small ventral bumps about half-way down the male tail, each with a pair of short setae. A large double-tipped papilla in front of cloaca, with six small rounded papillae along its posterior edge.

Distribution. Exe estuary; Isles of Scilly (coarse intertidal sand near low water).

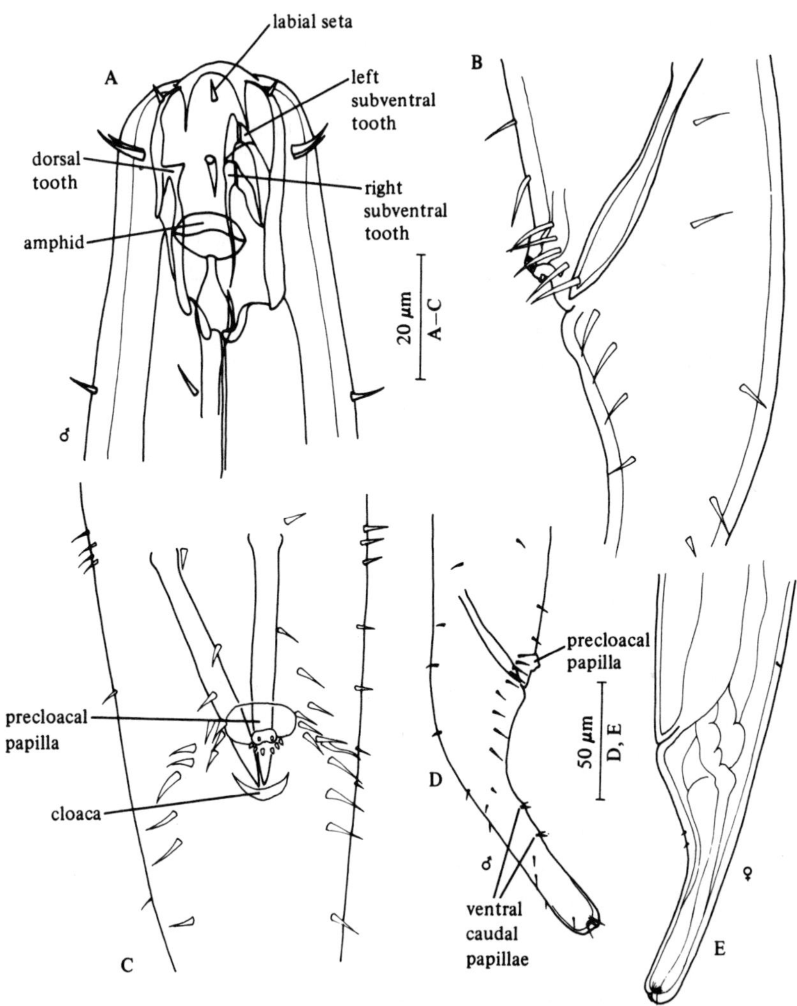

Fig. 102. *Oncholaimus campylocercoides*. A, Male head; B, Cloacal region (lateral); C, Cloacal region (ventral); D, Male tail; E, Female tail. Original.

Oncholaimus skawensis Ditlevsen, 1921
(Fig. 103)

Description. Body length 2.5–3.4 mm. Maximum diameter 56–103 μm (a = 33–45). Six small rounded labial papillae. Six longer cephalic setae 7–8 μm (0.3 h.d.), shorter four 4–5 μm. Left subventral tooth longer than other two. Amphids 12 μm wide (0.4 times c.d.). Excretory pore 1.6 buccal cavity lengths from anterior in male, just behind base of buccal cavity in female. Male tail 3.2 a.b.d. long, strongly constricted just behind cloaca with long cylindrical portion. Female tail 3.0 a.b.d., more evenly tapered.

Spicules straight, slender, 38–41 μm (1.9 a.b.d.).

No gubernaculum. About ten pairs of small rounded circumcloacal papillae. Two raised ventral bumps on the male tail each with a pair of setae.

Distribution. Northumberland (sublittoral mud).

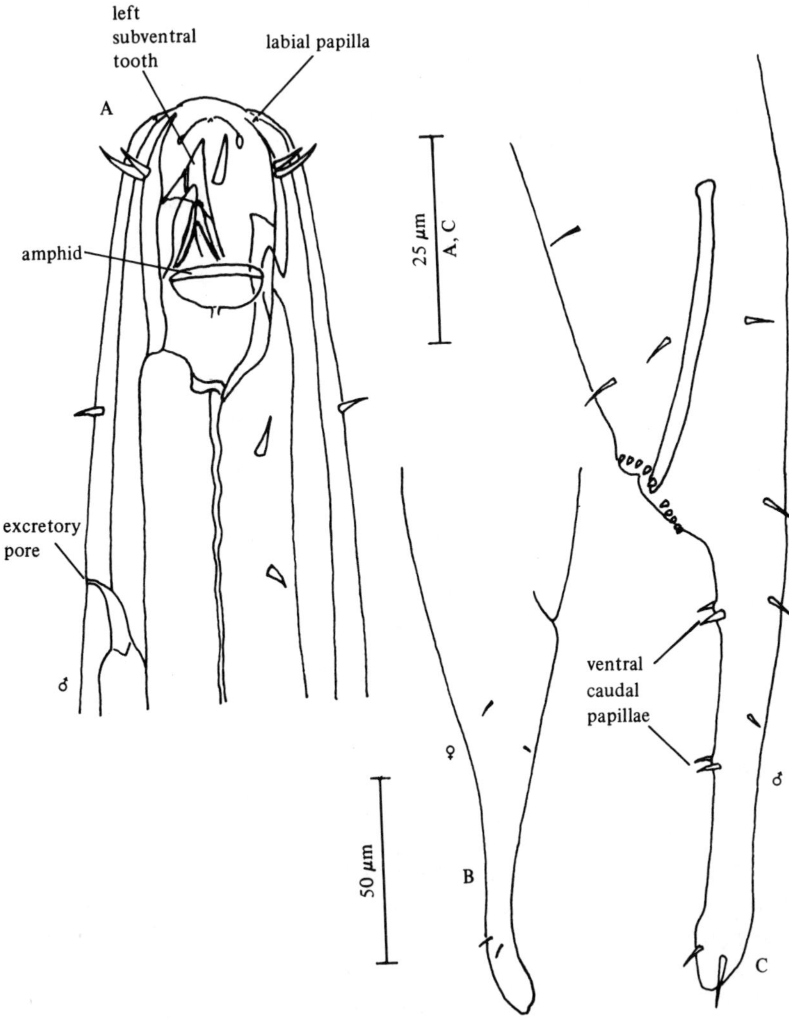

Fig. 103. *Oncholaimus skawensis*. A, Male head; B, Female tail; C, Male tail. Original.

Oncholaimus attenuatus Dujardin, 1845
(Fig. 104)

Description. Body length 2.8 mm. Maximum diameter 60 μm (a = 47). Very similar to *O. skawensis* but spicules shorter (1.6 a.b.d.) and stouter and seven pairs of fairly long and slender circumcloacal setae.

Distribution. Falmouth; Clare Island, West Ireland; Whitstable (intertidal sediments).

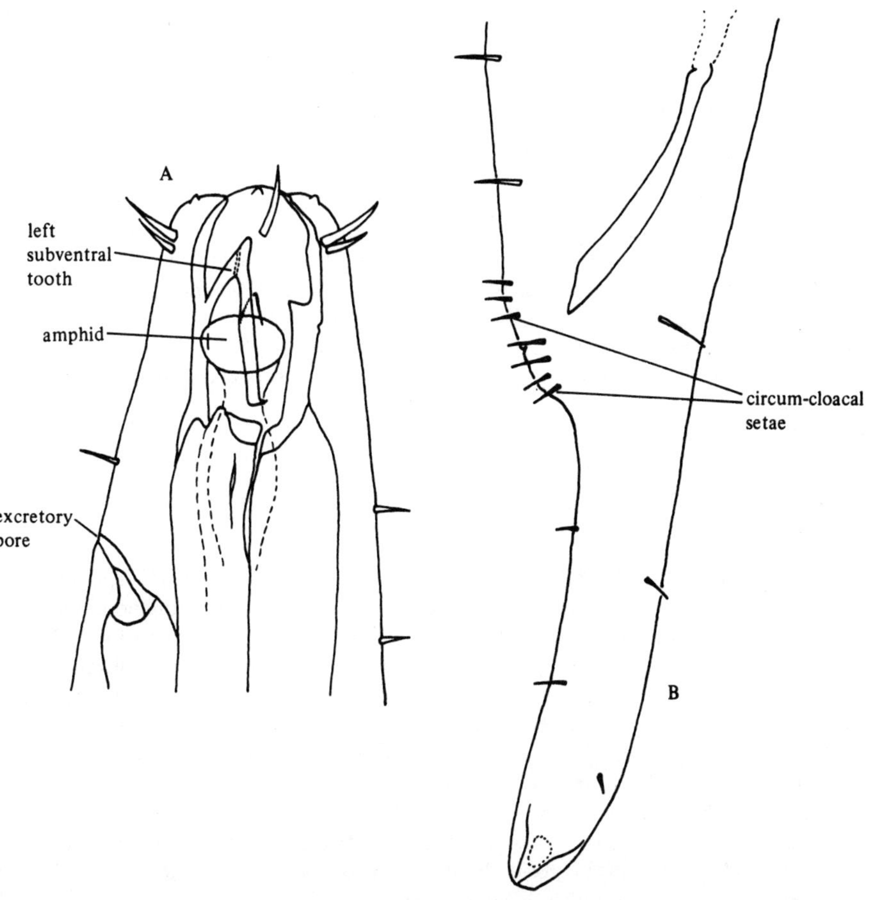

Fig. 104. *Oncholaimus attenuatus*. A, Male head; B, Male tail (from Stekhoven, 1931).

Genus PONTONEMA Leidy, 1855

Key: Mawson, 1956
Species of *Pontonema* are some of the largest, most commonly found marine nematodes. The characteristic feature of the genus is the equal-sized subventral teeth extending anterior to the dorsal tooth. The three species known so far from British waters can be distinguished mainly on the relative sizes of dorsal and subventral teeth, position of the excretory pore and male copulatory structures. The two species described by Southern (1914) have never been found since, but some of the original specimens are available. Southern stated that, on the west coast of Ireland, *P. vulgare* was almost invariably found with *P. simile*, which closely resemble each other and are only distinguished by the male copulatory structures.

The key given by Mawson (1956) contains a few errors and should be used with care.

Species : 29

Pontonema vulgare (Bastian, 1865)
(Fig. 105)

Description. Body length 12.3–14.9 mm. Maximum diameter 170–220 μm (a = 65–79). Cuticle smooth. Short somatic setae present throughout the body. Six short labial setae. Ten cephalic setae: six 10–12 μm (0.14–0.17 h.d.) and four 8–9 μm (0.11–0.13 h.d.). Amphids situated a short distance behind lateral cephalic seta, about level with dorsal tooth; 11–13 μm wide (0.15–0.17 c.d.). Buccal cavity large, about 1.6–1.9 times as long as wide. Two subventral teeth of similar size and extending further anterior than dorsal tooth. Excretory pore about 4.2–4.9 buccal cavity lengths from anterior. Nerve ring 0.36–0.39 times oesophagus length from anterior. Oesophagus length 0.09–0.12 times body length, gradually widening throughout its length but no posterior bulb. Tail very short, 145–155 μm (1.3–1.5 a.b.d.); rounded with well developed spinneret.

Spicules 120–140 μm (1.1–1.2 a.b.d.), slender.

Gubernaculum 50–60 μm long. Ventral papilla with spines present just anterior to cloaca and another 195–220 μm (1.8–2.0 a.b.d.) anterior to cloaca. Two opposed testes.

Two reflexed ovaries. Vulva at 53–56% of body length.

Distribution. Falmouth (intertidal sand and stones); St Andrews, Scotland (intertidal under stones); West Ireland (intertidal); Plymouth (intertidal among weeds); Northumberland (kelp holdfasts); Isles of Scilly (kelp holdfasts and other weeds); Strangford Lough, Northern Ireland (among weed). Very common and widespread.

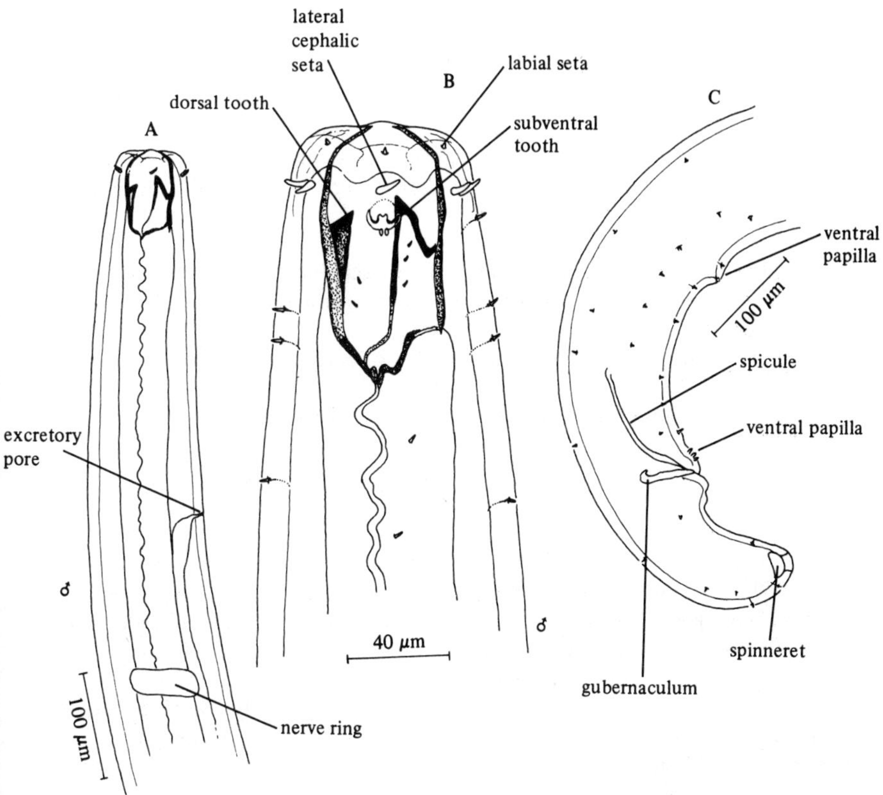

Fig. 105. *Pontonema vulgare*. A, Anterior end of male; B, Male head; C, Posterior end of male. Original.

Pontonema macrolaimus (Southern, 1914)
(Fig. 106A, B, C)

Description. Body length 8.9–12.6 mm. Maximum diameter 154–186 μm (a = 55–68). Amphid large. Dorsal tooth far behind the subventral teeth. Excretory pore opens just posterior to buccal cavity. Nerve ring 0.35–0.37 times oesophagus length from anterior. Tail short and bluntly conical.

Spicules 100 μm (0.9 a.b.d.).

Gubernaculum present. Posterior to cloaca there are numerous small spines. Just anterior to cloaca is a papilla and a second small papilla is situated 124 μm (1.1 a.b.d.) from cloaca.

Two reflexed ovaries. Vulva at 54–58% of body length.

Distribution. West Ireland (sublittoral in sand and shells).

Pontonema simile (Southern, 1914)
(Fig. 106D, E, F)

Description. Body length 11.3, 18.0 mm. Maximum diameter 214, 280 μm (a = 53, 64). Buccal cavity similar to *P. vulgare.* Excretory pore about 4 buccal cavity lengths from anterior. Nerve ring about 0.35 times oesophagus length from anterior. Tail short and rounded.

Spicules large and stout, 184 μm (1.5 a.b.d.).

Gubernaculum present with a conspicuous tooth distally. Papillae present just anterior to cloaca and anterior to proximal end of spicule (about 1.5 a.b.d. from cloaca).

Two reflexed ovaries. Vulva at 55% of body length.

Distribution. West Ireland (intertidal).

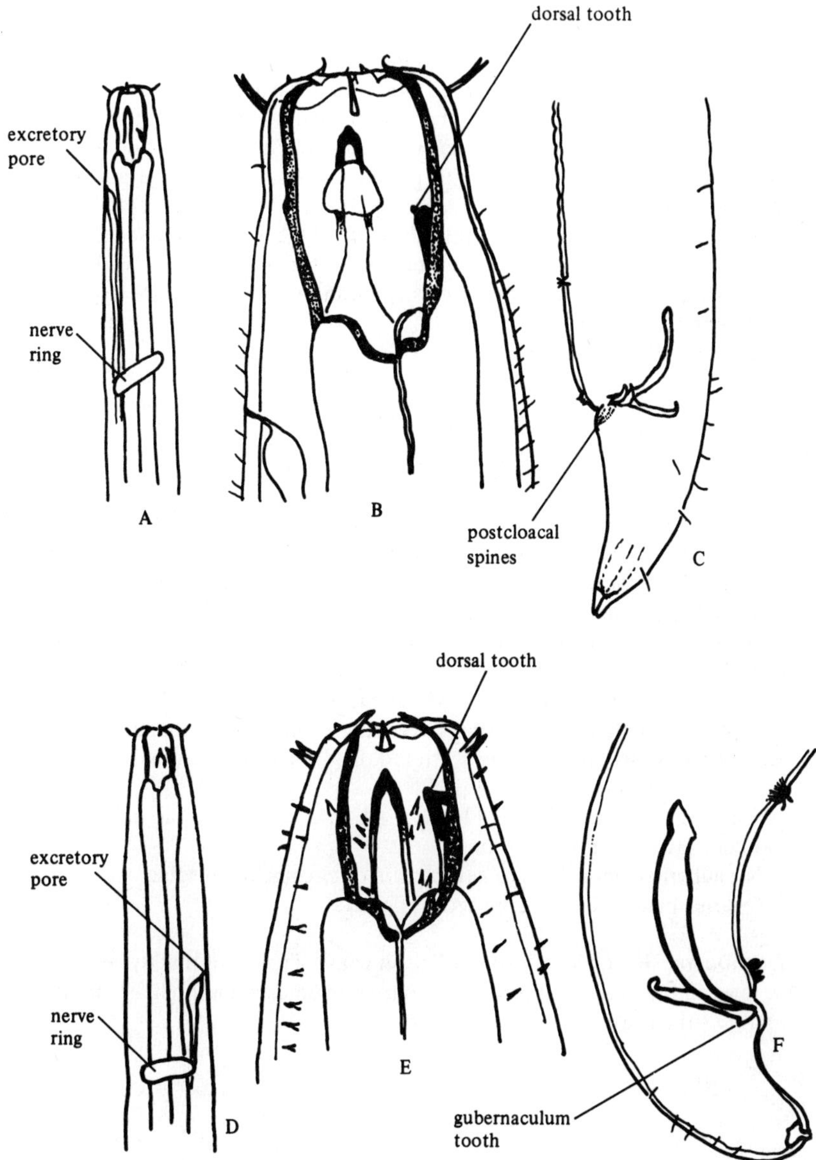

Fig. 106. *Pontonema macrolaimus*. A, Anterior end; B, Head; C, Posterior end of male (from Southern, 1914). *Pontonema simile*. D, Anterior end; E, Head; F, Posterior end of male (from Southern, 1914).

Genus VISCOSIA De Man, 1890

Keys: Wieser, 1953; Wieser and Hopper, 1967
The genus *Viscosia* is distinguished from other oncholaimids by a combination of three subequal teeth in the buccal cavity with the right subventral larger than the others, short spicules, no gubernaculum, male with no copulatory bursa and female with paired ovaries. It is a large and rather difficult genus, the species being separated on the length of the cephalic setae or papillae, the size of the amphids, the form of the dorsal and left subventral teeth in the buccal cavity, the length of the tail and (as with other oncholaimids) the male copulatory apparatus. All species are typical of sedimentary habitats.

Species : 60

Viscosia viscosa (Bastian, 1865)
(Fig. 107)

Description. Body length 1.7–2.0 mm. Maximum diameter 30–36 μm (a = 52–60). Cuticle smooth. Six rounded lips with six small rounded labial papillae. Six longer cephalic setae 4 μm (0.3 h.d.), remaining four slightly shorter. Head constricted just posterior to cephalic setae. Buccal cavity with large right subventral tooth, left subventral tooth double tipped, dorsal tooth small, single tipped. Amphids 9–10 μm wide (male), 8 μm (female), a little over half the c.d., pocket-like with elliptical openings. Oesophagus 0.17 times body length, more or less cylindrical. Excretory pore about 0.7 times oesophagus length from anterior, nerve ring at 0.5 times oesophagus length. Body setae short and sparsely scattered. Tail 6 a.b.d., proximal two-thirds conical, distal third cylindrical.

Spicules equal, 27–30 μm (1.3–1.5 a.b.d.), curved, distal half slightly swollen, proximal tip cephalate.

No gubernaculum. Five to seven pairs of circumcloacal setae.

Ovaries paired, equal, opposed, reflexed.

Distribution. Recorded all round British coast. Typical of muddy sediments in low salinity areas, but also from coarser sediments and holdfasts in more marine situations.

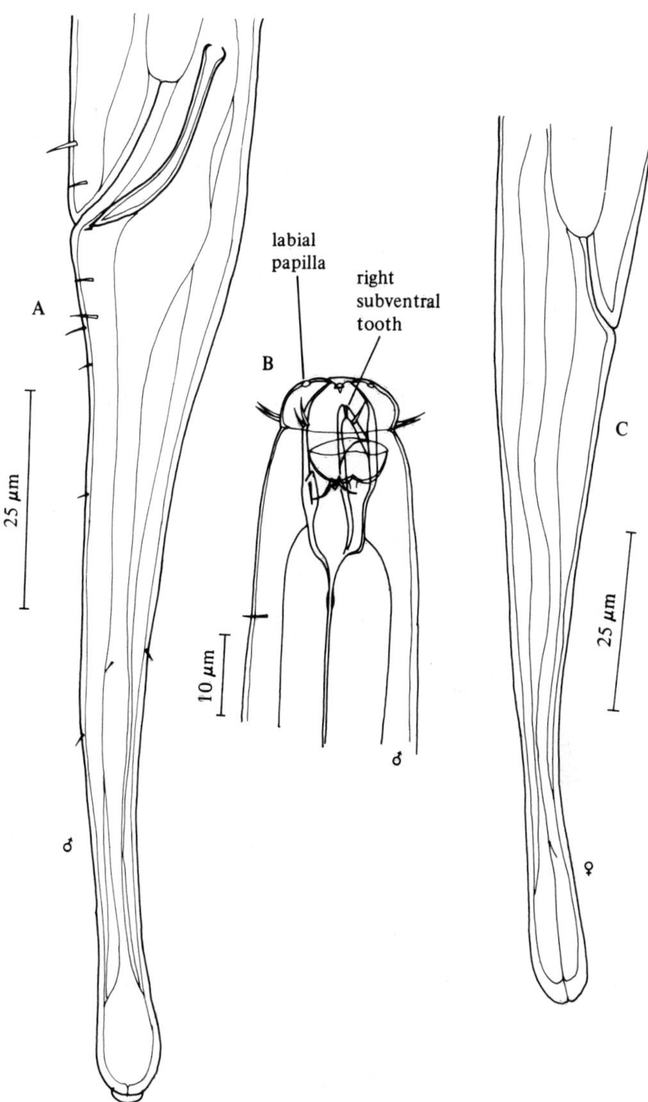

Fig. 107. *Viscosia viscosa*. A, Male tail; B, Male head; C, Female tail. Original.

Viscosia abyssorum (Allgén, 1933)
(Fig. 108)

Description. Relatively short and stout compared with other species. Body length 1.8–2.0 mm. Maximum diameter 50–56 μm (a = 35). Labial papillae minute, scarcely visible. Six longer cephalic setae 4.5 μm (0.3 h.d.), other four slightly shorter. Right subventral tooth large, pointed, other two teeth weakly developed, round tipped. Amphids 10.5 μm wide (half the c.d.). Tail relatively short (3.3–3.8 a.b.d.), with distinctly swollen tip.

Spicules straight, slender, 26–32 μm (1.1 a.b.d.).

Five pairs of circumcloacal setae in male, with paired setae extending down tail.

Distribution. Skippers Island, Essex (muddy intertidal); Northumberland (sublittoral mud).

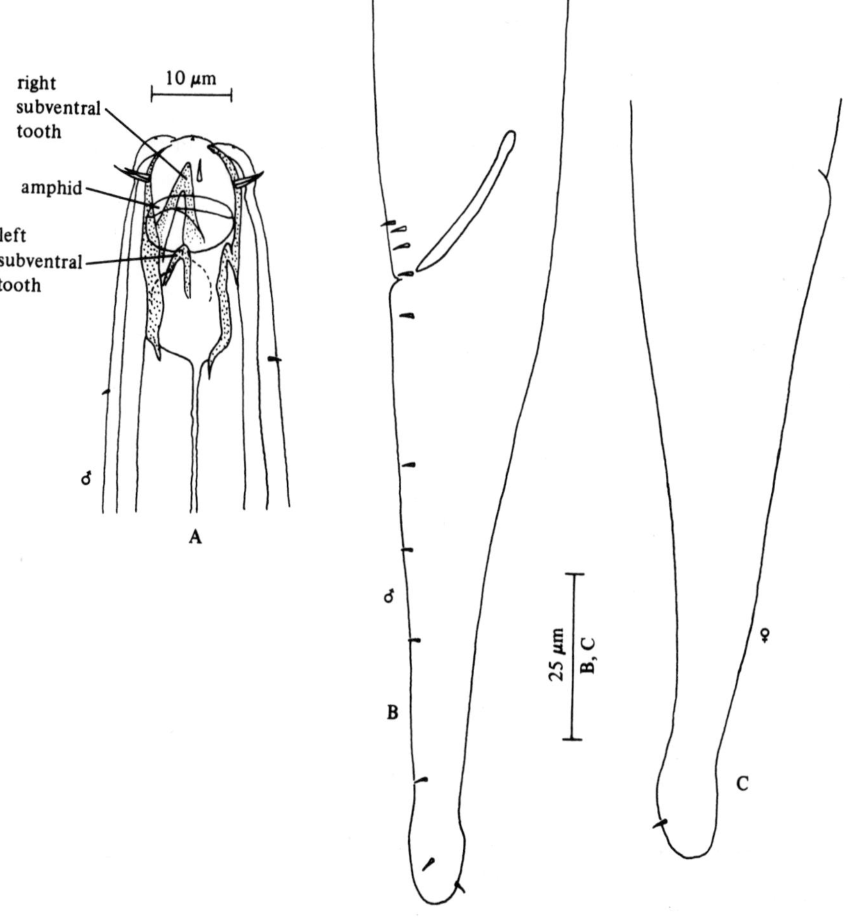

Fig. 108. *Viscosia abyssorum*. A, Male head; B, Male tail; C, Female tail. Original.

Viscosia cobbi Filipjev, 1918
(Fig. 109)

Description. Body length 2.3–2.8 mm. Maximum diameter 33–45 μm (a = 62–78). Six short and four slightly longer (6–7 μm or one-third of h.d.) cephalic setae. Right subventral tooth larger than remaining two, which are both prominent and single tipped. Amphids 11 μm wide, 0.5 h.d. (male); 7 μm, 0.33 h.d. (female). Tail 7 a.b.d., anterior quarter tapering, posterior part cylindrical, or often appearing cylindrical throughout.

Spicules fairly straight, 22–27 μm (1 a.b.d.). Two pairs of setae in front of cloaca, three pairs behind, one small seta either side of cloaca.

Distribution. Exe estuary; Strangford Lough, Northern Ireland (intertidal sand).

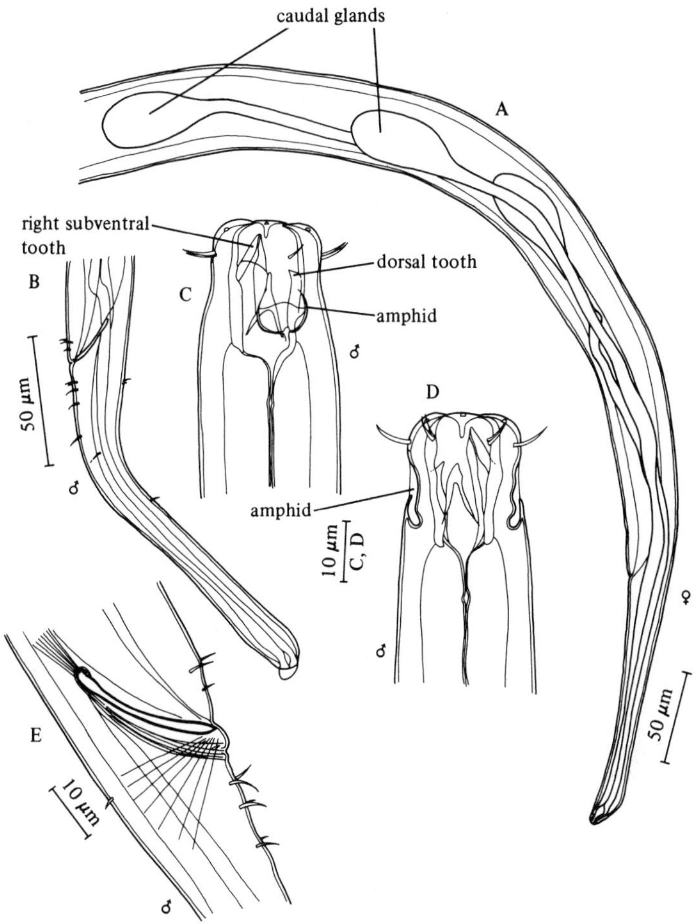

Fig. 109. *Viscosia cobbi.* A, Posterior end of female; B, Male tail; C, Male head (lateral); D, Male head (dorsal); E, Cloacal region. Original.

Viscosia glabra (Bastian, 1865)
(Fig. 110)

Description. Body length 1.9 mm. Maximum diameter 39 μm (a = 49). Cephalic sense organs represented by six small papillae only. Amphids 7–8 μm wide (0.41–0.46 times c.d.). Right subventral tooth large, other two single tipped, more or less equal. Tail elongate (10.5 a.b.d.).

Spicules 27 μm (1.4 a.b.d.), tip slightly bifurcate. Few very short circum-cloacal setae.

Distribution. Penzance; Isles of Scilly; Whitstable; Skippers Island, Essex; Northern Ireland (intertidal sand); North East England (*Laminaria* hold-fasts).

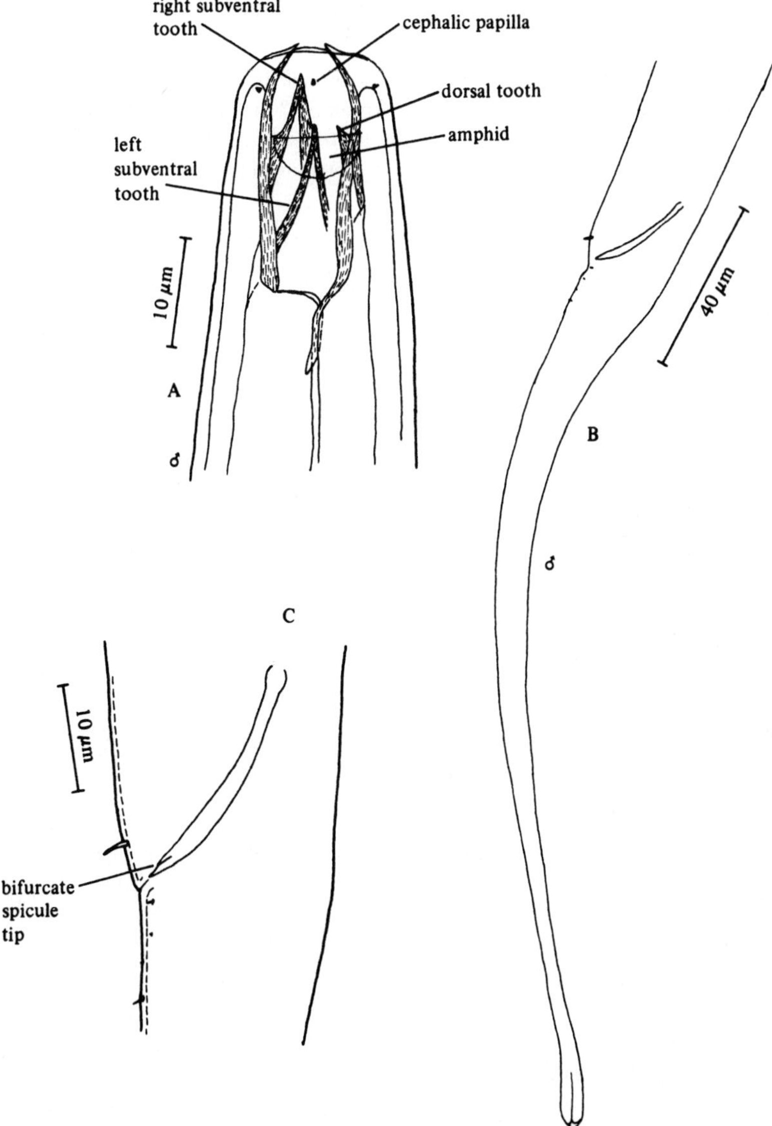

Fig. 110. *Viscosia glabra*. A, Male head; B, Male tail; C, Cloacal region. Original.

Viscosia langrunensis (De Man, 1890)
(Fig. 111)

Description. Body length 2.6 mm. Maximum diameter 26 μm (a = 100). Cephalic setae 3.5–4 μm (0.2 h.d.). Amphids 7 μm wide (0.37 times c.d.). Dorsal tooth and smaller subventral tooth extend anterior to middle of buccal cavity, both single tipped. Tail conico-cylindrical with slightly distended tip, 6.6 a.b.d.

Spicules 22 μm (1.3 a.b.d.).

Distribution. Northern Ireland (intertidal sand).

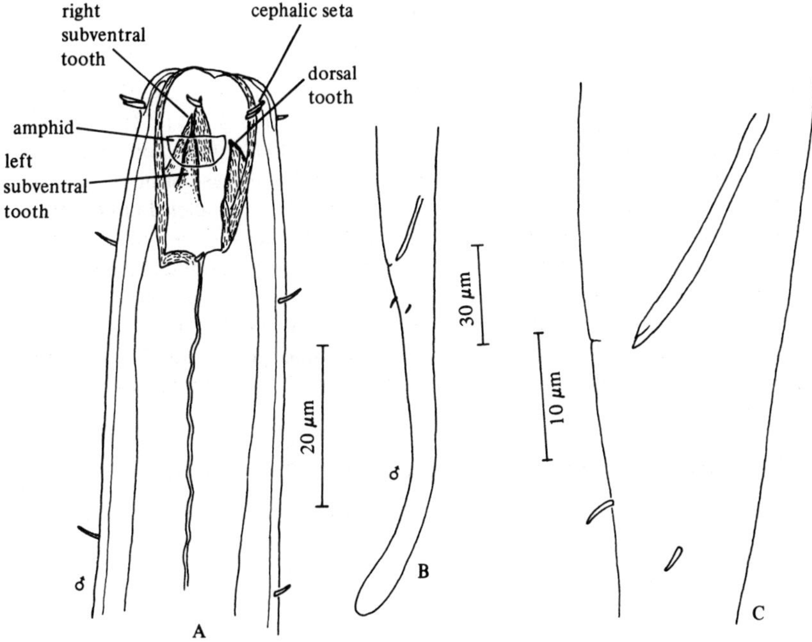

Fig. 111. *Viscosia langrunensis*. A, Male head; B, Male tail; C, Cloacal region. Original.

Viscosia elegans (Kreis, 1924)
(Fig. 112)

Description. Body length 2.0–3.1 mm. Maximum diameter 30–50 μm (a = 63–73). Six labial papillae. Six longer cephalic setae 3–4 μm (0.25 h.d.) and four slightly shorter; pairs sometimes stick together giving appearance of only six cephalic setae. Lateral cervical seta level with base of buccal cavity. Amphids 10 μm (0.6 times c.d.) wide. Large right and smaller double-tipped left subventral teeth; dorsal tooth represented by slight ridge only. Tail 6–9 a.b.d., gradually tapering, bulbous tip with subdorsal terminal setae.

Spicules 20–27 μm (1.2–1.3 a.b.d.), slightly curved, cephalate proximally. Cloaca surrounded by paired setae which continue down tail.

This species is rather similar to *V. viscosa*, although the tail of the latter is shorter, the spicules more bent and the dorsal tooth more strongly developed.

Distribution. Isles of Scilly (sublittoral sand, 12–27 m depth).

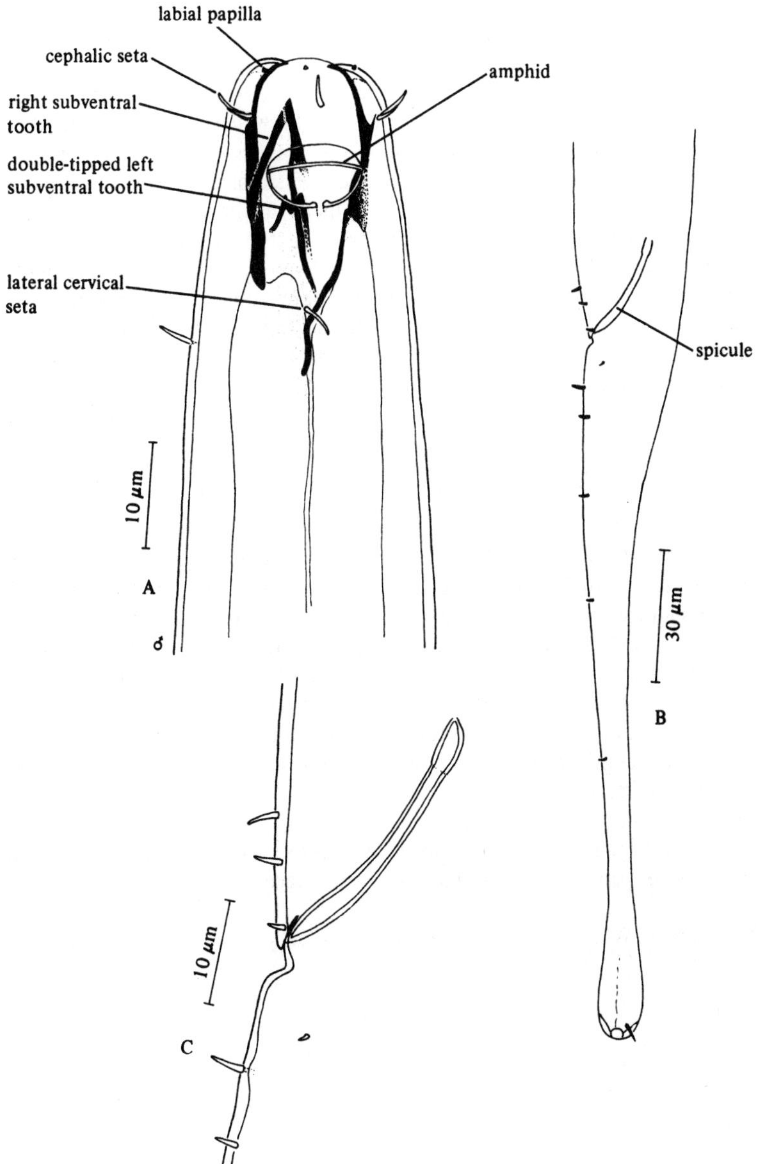

Fig. 112. *Viscosia elegans*. A, Male head; B, Male tail; C, Cloacal region. Original.

Genus BELBOLLA Andrassy, 1973

Key: Belogurov and Belogurova, 1980

Belbolla is a member of the family Enchelidiidae which contains relatively infrequently reported, often poorly known species which exhibit a high degree of intra-generic variation. Genera in this family usually have a divided buccal cavity and winged precloacal supplements. However, the characteristic feature of *Belbolla*, which distinguishes it from all other enoplids except *Polygastrophora*, is the presence of multiple oesophageal bulbs. *Polygastropha* differs from *Belbolla* in having more than two buccal cavity chambers and no supplements. However, one European species of *Belbolla* is known without supplements. The only male specimen known so far from British waters is that originally described by Inglis (1961) whose description is used here.

Species : 8

Belbolla gallanachmorae (Inglis, 1961)
(Fig. 113)

Description. Body length 2.9 mm. Maximum diameter 83 μm (a = 35). Head attenuated. Six small labial papillae. Ten cephalic setae in one circle of which six are slightly larger (10 μm) than the others. Pocket-like amphids dorsolateral in position, level with the middle of the buccal cavity. Buccal cavity 14 μm long and 5 μm wide, with a large right subventral tooth and two smaller left subventral and dorsal teeth and divided into two chambers by two transverse cuticular rings. Excretory pore and nerve ring 0.24 and 0.53 times oesophagus length from anterior respectively. Posterior part of oesophagus has eight bulbs. Tail 4.8 a.b.d., conico-cylindrical.

Spicules 69 μm (1.4 a.b.d.).

Short gubernaculum with dorsal apophysis. Two 'winged' precloacal supplements about 1.7 and 4.0 a.b.d. from cloaca.

Distribution. West Scotland (*Laminaria* holdfasts).

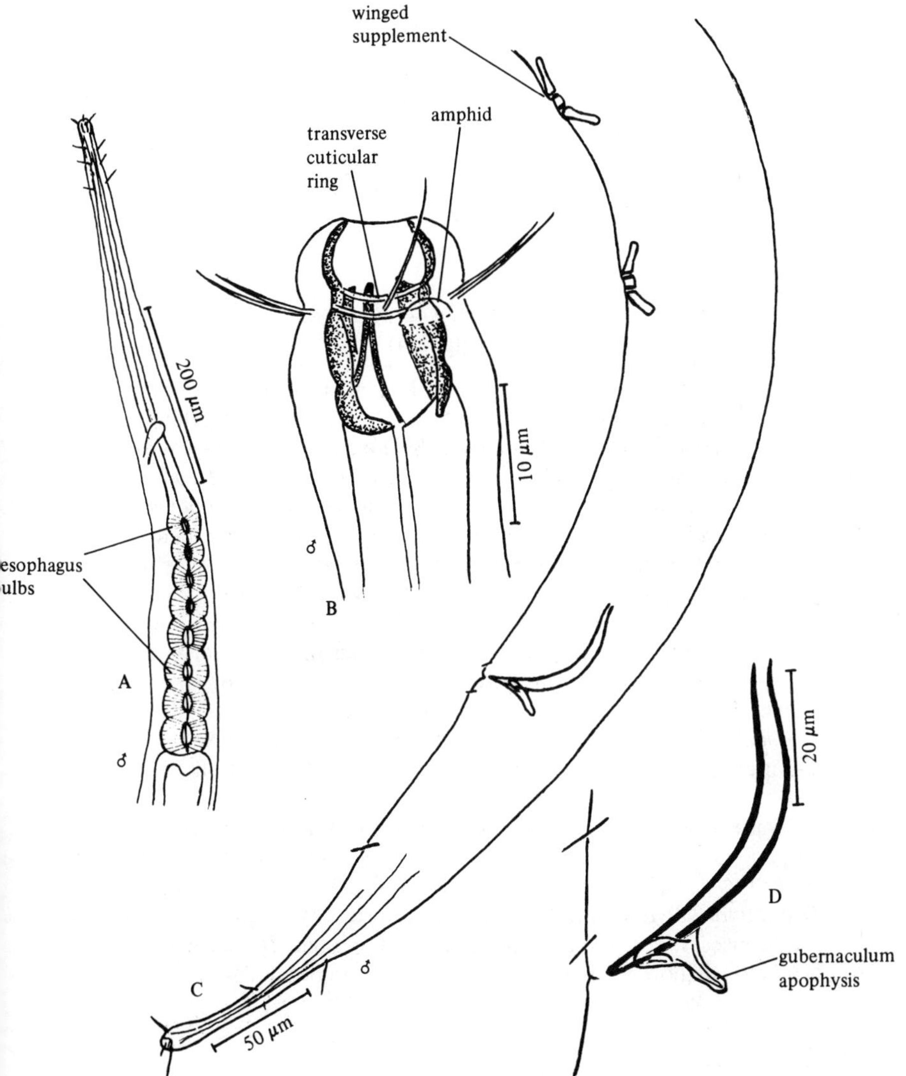

Fig. 113. *Belbolla gallanachmorae*. A, Anterior end of male; B, Male head; C, Posterior end of male; D, Spicules and gubernaculum (from Inglis, 1961).

Genus CALYPTRONEMA Marion, 1870

Key: Wieser, 1953
This genus exhibits strong sexual dimorphism. In the female the buccal cavity is typical of the Enchelidiidae consisting of two chambers, *the posterior one extending down the oesophageal lumen* to some distance behind the nerve ring. In the male there is no strongly cuticularised buccal cavity. The precloacal supplements are not winged.

Species : 16

Calyptronema maxweberi (De Man, 1922)
(Fig. 114)

Description. Body length 1.8–2.8 mm. Maximum diameter 49–67 μm (a = 29–45). Cuticle smooth. Head constricted a short distance behind cephalic setae. Six small conical labial papillae. Six longer cephalic setae 4 μm, other four slightly shorter. Buccal cavity in female tubular, divided by a fine suture, confluent with a conspicuous, wide, heavily cuticularised oesophageal lumen which extends just posterior to the nerve ring. Three slender teeth with one larger than the others; the latter can be protruded some distance out of the buccal cavity. Amphids lateral, pocket-like with elliptical openings, 4 μm wide. Buccal cavity in male absent; amphids relatively larger. Oesophagus 0.16–0.23 times body length, broadening gradually posteriorly. Excretory pore at level of cephalic constriction. Nerve ring half-way down oesophagus length. Tail 7 a.b.d. long, with a long cylindrical distal part.

Spicules 70–102 μm long, arcuate. A pair of stout setae anterior to cloaca; 4–6 precloacal supplements, the posterior two different from the anterior 2–4.

Ovaries paired, equal, opposed, reflexed. Vulva at 50–58% of body length.

Distribution. Plymouth (intertidal algae); North East England (holdfasts and rock crevices); Exe estuary (intertidal mud and muddy sand); Loch Etive, Scotland (shallow sediments); Isles of Scilly (intertidal algae).

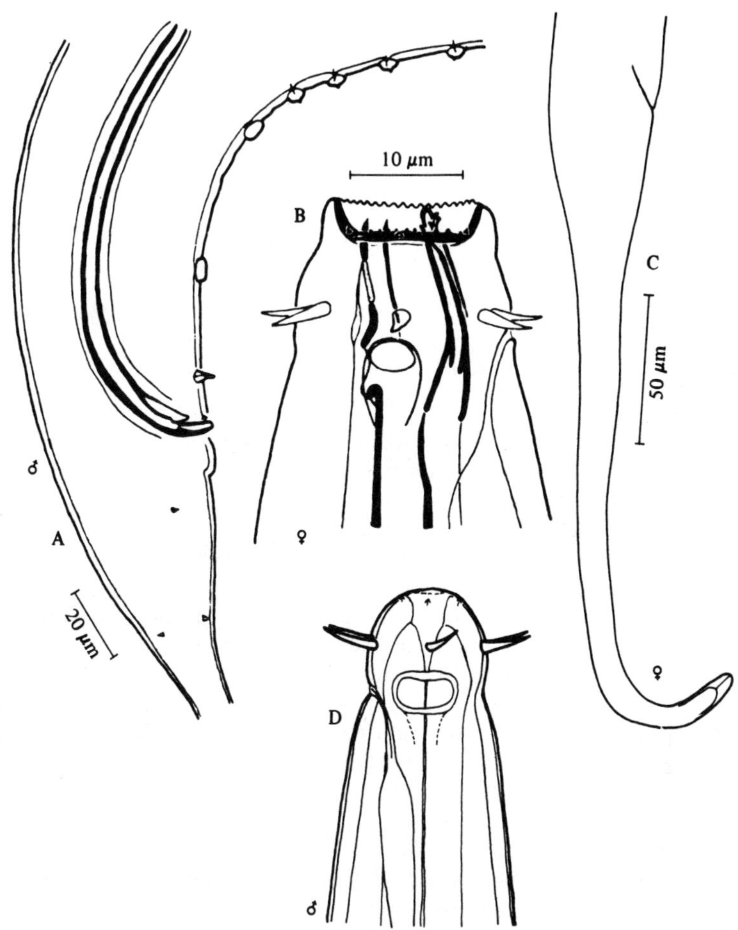

Fig. 114. *Calyptronema maxweberi*. A, Cloacal region; B, Female head (from Lorenzen, 1969); C, Female tail (original); D, Male head (from Bresslau and Stekhoven, 1940).

Genus EURYSTOMINA Filipjev, 1921

Key: Wieser, 1953

Eurystomina is the most well known genus of the Enchelidiidae, containing forms with rows of buccal denticles, cuticularised supplements and functional caudal glands. The only key available (Wieser, 1953) is now somewhat out-of-date, containing only eight of the eighteen valid species. The *E. ornata* from the British Isles were originally reported as *E. filiformis* but this latter species, together with two others, were all recently synonymised (Lambshead and Platt, 1979) since all four are similar although poorly known.

Species : 18

Eurystomina ornata (Eberth, 1863)
(Fig. 115)

Description. The following description is based on a male (first measurement) and a female from the North Sea.

Body length 3.5, 4.3 mm. Maximum diameter 44, 58 μm (a = 79, 75). Cuticle smooth. Short but conspicuous cervical setae anterior to nerve ring, otherwise somatic setae absent. Six labial papillae. Ten cephalic setae in one circle; six 7.5, 11 μm (0.4, 0.5 h.d.) and four 5, 6 μm (0.3, 0.3 h.d.). Amphids a reniform loop, width about one-third of c.d. and lying dorso-sublaterally. Ocelli present, lying 1.6 h.d. from anterior, consisting of orange granules. Buccal cavity divided into two sections by rows of denticles; two distinct fairly regular rows in both sexes plus an extra more irregular row in the female. Large right subventral tooth but other teeth inconspicuous. Excretory pore opens level with the amphid. Nerve ring 0.4 times oesophagus length from anterior. Tail conical, 2.2, 4.1 a.b.d. Caudal glands extend anterior to anus.

Spicules 120 μm (arc).

Gubernaculum 38 μm. Two cuticularised winged supplements 1.7 and 2.9 a.b.d. from cloaca. Two opposed testes.

Two opposed reflexed ovaries. Vulva at 57% of body length.

Distribution. West Ireland (intertidal amongst *Spirorbis* tubes and sublittoral in sand and shells); Plymouth (intertidal weed); Northumberland (kelp holdfasts); Isles of Scilly (kelp holdfasts and amongst other weeds).

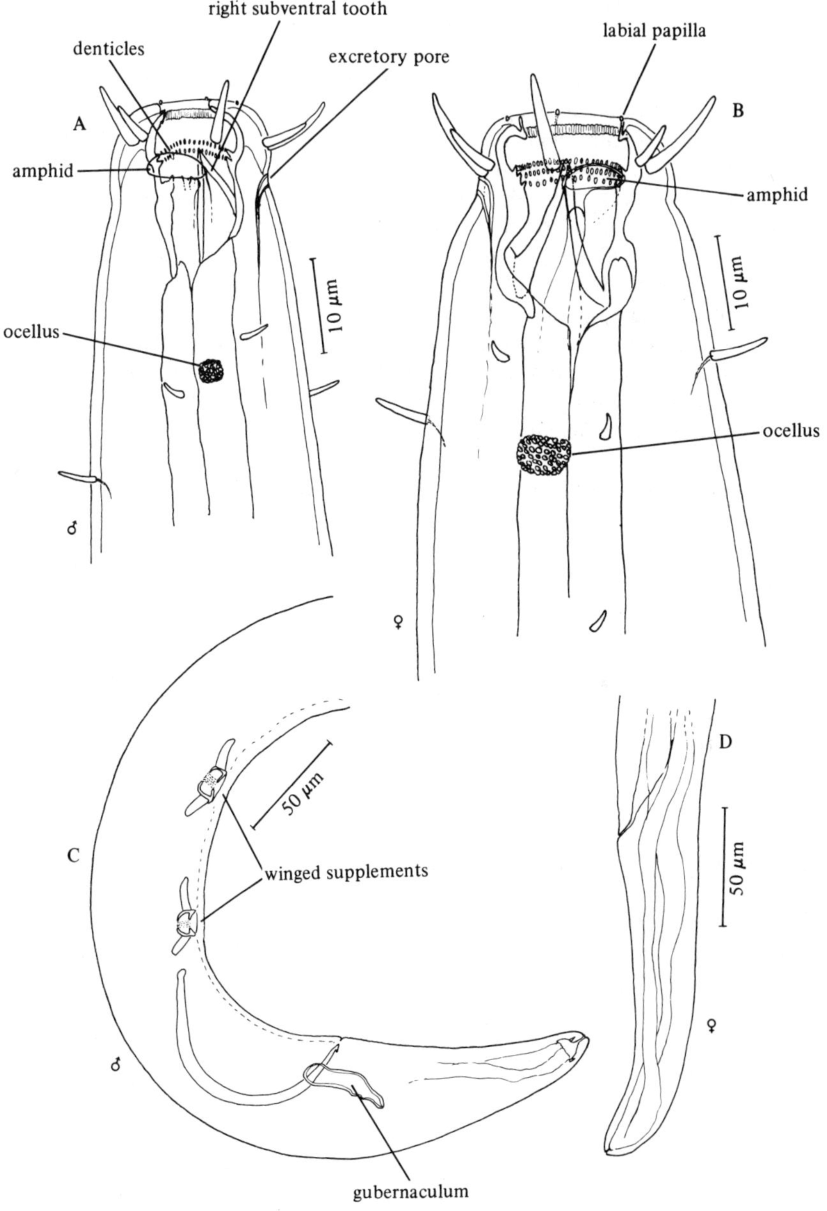

Fig. 115. *Eurystomina ornata*. A, Male head; B, Female head; C, Posterior end of male; D, Female tail. Original.

Eurystomina cassiterides (Warwick, 1977)
(Fig. 116)

Description. Body length 3.5 mm. Maximum diameter 51 μm (a = 69). Cuticle smooth. Some longish cervical setae and short setae scattered sparsely on tail. Six labial setae relatively long (5 μm) and slender. Ten cephalic setae in one circle; six 9 μm and four 10 μm. Amphids pocket-like, 9 μm wide. Buccal cavity divided into two sections by band of denticles. Large right subventral tooth and smaller left subventral and dorsal teeth. Excretory pore and nerve ring 0.05 and 0.28 times oesophagus length from anterior respectively. Tail 112 μm (2.8 a.b.d.), conico-cylindrical.

Spicules arcuate, 45 μm (1.1 a.b.d.).

Gubernaculum with 30 μm apophysis. Five precloacal supplements; the posterior three simple ventral swellings 0.3, 0.8 and 2.2 a.b.d. from cloaca, progressively increasing in size; anterior two 3.8 and 5.8 a.b.d. from cloaca each with minute cuticularized 'wings'.

Distribution. Isles of Scilly (sublittoral shell gravel at 52 m).

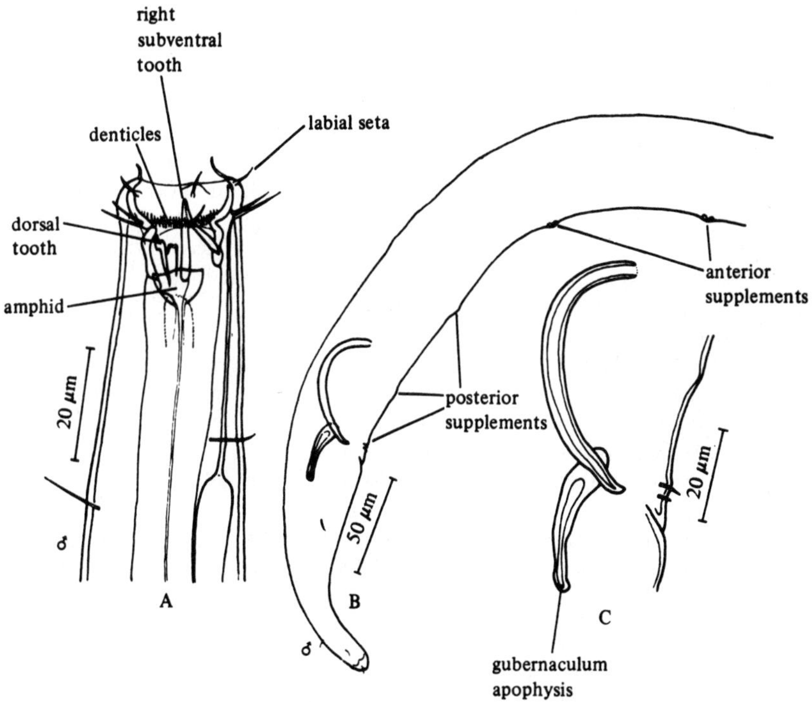

Fig. 116. *Eurystomina caesiterides*. A, Male head; B, Posterior end of male; C, Spicules and gubernaculum. Original.

Eurystomina terricola (De Man, 1907)
(Fig. 117)

Description. Body length 3.3–4.4 mm. Maximum diameter 58–64 μm (a = 56–68). Cuticle smooth. Six labial papillae. Ten cephalic setae in one circle; six 10–12 μm (0.4 h.d.) and four 4–6 μm (0.2 h.d.). Amphids dorso-sublateral in position and about 0.4 h.d. wide. Ocelli absent. Buccal cavity divided into chambers by three rows of denticles. Large right subventral and smaller more rounded dorsal and left subventral teeth. Excretory pore about 1.5 buccal cavity lengths from anterior. Nerve ring about 0.35 times oesophagus length from anterior. Tail conical, 2.3 a.b.d. Caudal glands extend anterior to anus.

Spicules 98–104 μm.

Gubernaculum 31–35 μm. Two cup-shaped cuticularised supplements without wings 2.7 and 4.0–4.5 a.b.d. from cloaca; anterior larger than posterior. Two opposed testes.

Two opposed reflexed ovaries. Vulva at 62–3% of body length.

Distribution. Exe estuary (extreme upper level of sandy shore); Northumberland (dryer crevices on rocky shore); Isles of Scilly (rotting seaweed on strandline).

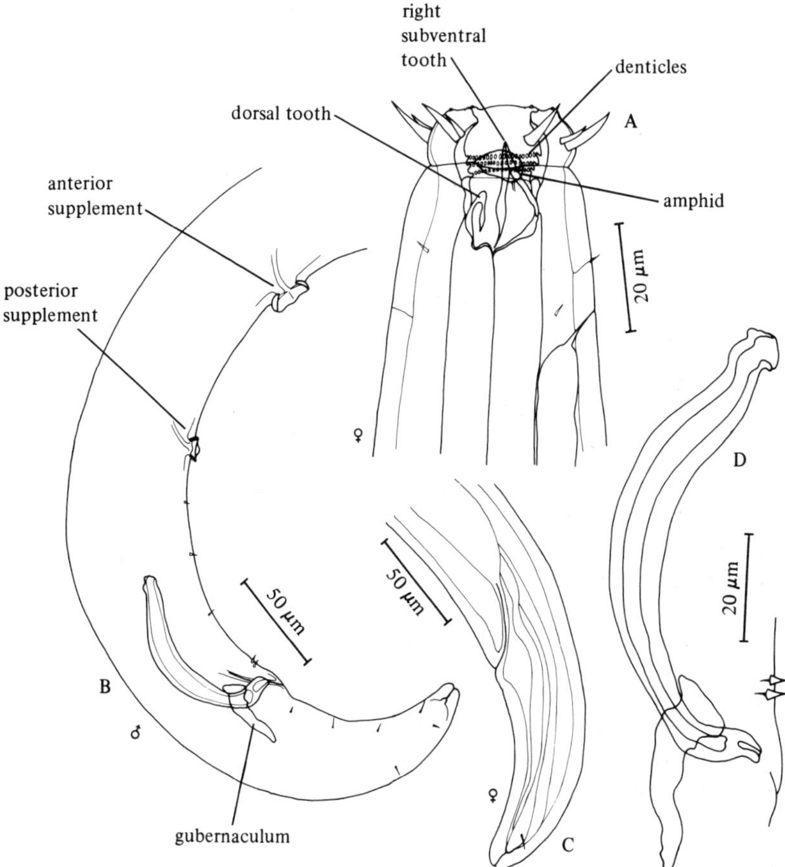

Fig. 117. *Eurystomina terricola*. A, Female head; B, Posterior end of male; C, Female tail; D, Spicules and gubernaculum. Original.

Genus PAREURYSTOMINA Micoletzky, 1930

Keys: Luc and De Coninck, 1959; Wieser, 1959

Pareurystomina has a head similar to that of *Eurystomina*, but is very characteristic in having a finely pointed tail with no caudal glands or spinneret. The two species so far reported from the British Isles can be separated on many characters (head, tail length and copulatory apparatus in the male).

Species : 14

Pareurystomina scilloniensis Warwick, 1977
(Fig. 118)

Description. Body length 7.1–7.8 mm. Maximum diameter 71–80 μm (a = 88–110). Cuticle smooth. Six labial papillae small, rounded. Cephalic setae 16 and 6.5 μm (male), 19 and 7.5 μm (female). Buccal cavity with single right subventral tooth slender, pointed; two complete circles of large denticles and an incomplete circle of very fine denticles. Amphids 10–12 μm wide, dorsolateral. Oesophagus 0.16 times body length, no posterior bulb. Excretory pore opens at level of cephalic setae. *Flattened cervical setae* extend posteriorly as far as the nerve ring. Tail tapering to a fine point, 4 a.b.d. (male), 6 a.b.d. (female).

Spicules arcuate, hooked distally, 82 μm long.

Gubernaculum dorsally directed, solid, 37 μm. Male tail with a pair of subventral double-pointed papillae about half-way down its length. Two winged supplements about 4 and 7 a.b.d. in front of cloaca, wings small and symmetrical. A small raised area of cuticle pierced by a fine pore about 1.5 a.b.d. in front of cloaca. A pair of short setae immediately anterior to cloaca.

Vulva at 69% of body length.

Distribution. Isles of Scilly (fine sublittoral sand).

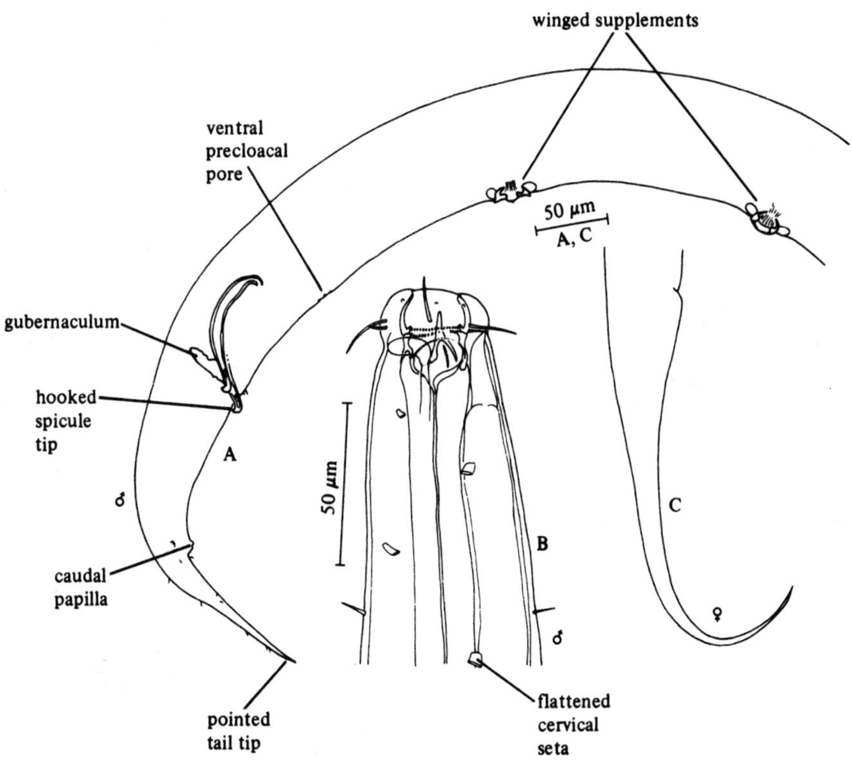

Fig. 118. *Pareurystomina scilloniensis*. A, Posterior end of male; B, Male head; C, Female tail. Original.

Pareurystomina acuminata (De Man, 1889)
(Fig. 119)

Description. Body length 4.2–4.5 mm. Maximum diameter 53–63 μm (a = 67–90). Longer cephalic setae 24 μm (male) and 17 μm (female). Amphids 30 μm wide (male) and 17 μm (female). Slender dorsal tooth and two complete circles of denticles in the buccal cavity. Tail finely pointed, 3.0–3.7 a.b.d. long.

Spicules 77 μm long. Gubernaculum with strong caudally directed apophysis. Precloacal supplements with long, paired, wing-like apophyses. No ventral papillae on tail.

Distribution. Penzance; Exe estuary; West Ireland (intertidal and shallow subtidal sand).

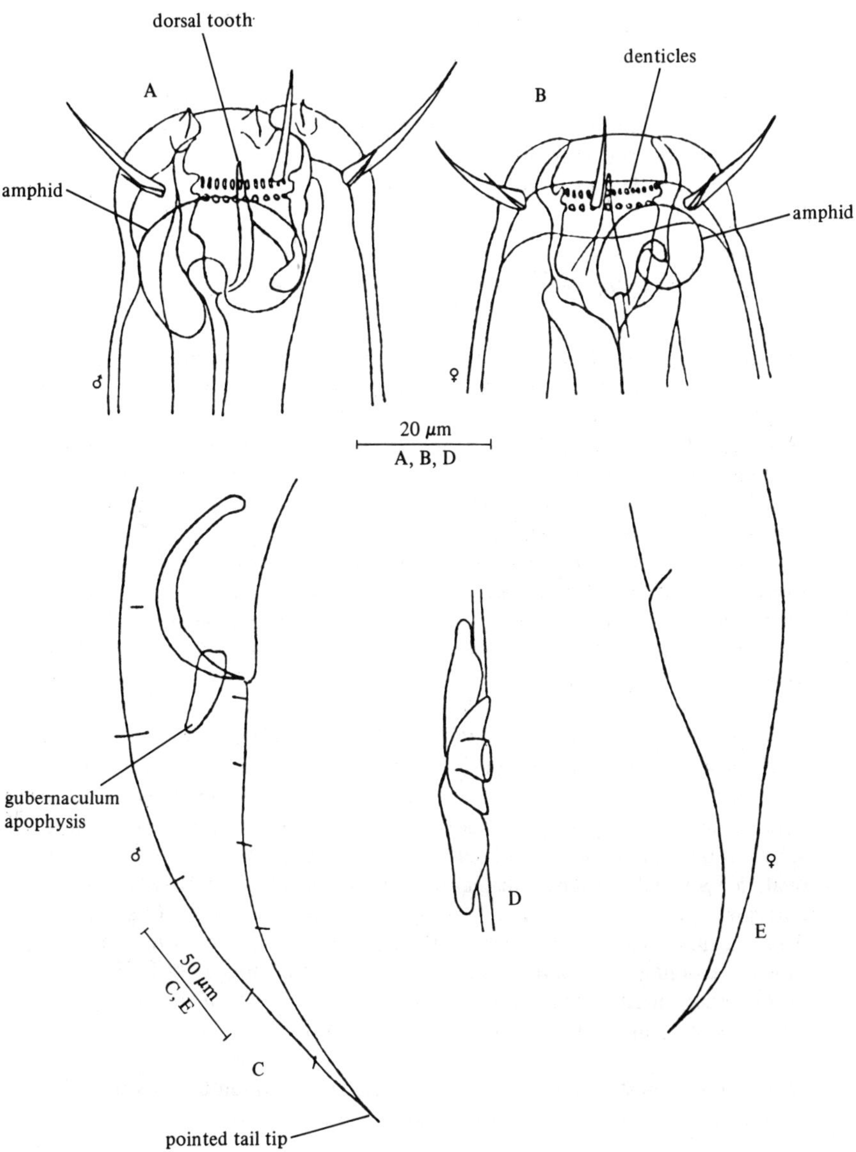

Fig. 119. *Pareurystomina acuminata*. A, Male head; B, Female head; C, Male tail; D, Precloacal supplement; E, Female tail (from Gerlach, 1952).

Genus SYMPLOCOSTOMA Bastian, 1865

Key: Wieser, 1953
This genus, like *Calyptronema*, has strong sexual dimorphism. The female has the typical chambered buccal cavity of the Enchelidiidae but in *Symplocostoma* the posterior chamber is not continued down the oesophageal lumen. The male lacks a buccal cavity.

Species : 15

Symplocostoma tenuicolle (Eberth, 1863)
(Fig. 120)

Description of male. Body length 4.2–7.0 mm. Maximum diameter 55–60 μm (a = 77–117). Cuticle smooth. Body strongly tapered at both ends. Head end slightly swollen, with a marked constriction behind the cephalic setae. Buccal cavity absent. Six small labial papillae. Cephalic setae 12 and 4 μm. Amphids slit-like, 4 μm wide, just anterior to the cephalic constriction. Large pigmented eyespot with paired ocelli 15 μm from anterior. Oesophagus 0.12–0.19 times body length, expanding posteriorly. Excretory pore 0.12 times oesophagus length from anterior; nerve ring 0.57 times oesophagus length from anterior. Spicules curved, 160 μm long (4.4 a.b.d.). Gubernaculum small, 40–51 μm (1.1–1.2 a.b.d.). 38–47 precloacal supplements (simple ventral swellings) extending 615 μm from cloaca. Tail conical, 4.8–4.9 a.b.d. long.

Description of female. Body length 5.1–6.1 mm. Maximum diameter 76–115 μm (a = 45–80). Cuticle smooth. Body strongly attenuated anteriorly, less so posteriorly. Head not swollen; no cephalic constriction. Six cephalic setae 5 μm long. Buccal cavity cylindrical, heavily cuticularised, with a large subventral tooth and a circle of denticles near the tip of this tooth. Amphids oval, 5.5 μm wide, slightly displaced dorsally. Oval ocelli 16–18 μm from anterior, eyespot smaller and less pigmented than in male. Oesophagus 0.19–0.2 times body length, expanded posteriorly. Excretory pore 0.08–0.09 times oesophagus length from anterior; nerve ring 0.48–0.53 times oesophagus length from anterior. Ovaries paired; vulva at 41–55% of body length. Tail evenly tapered, 4.5–6.1 a.b.d. long.

Distribution. West Ireland; North East England; Plymouth; Isles of Scilly (intertidal and subtidal seaweeds, holdfasts and polyzoa).

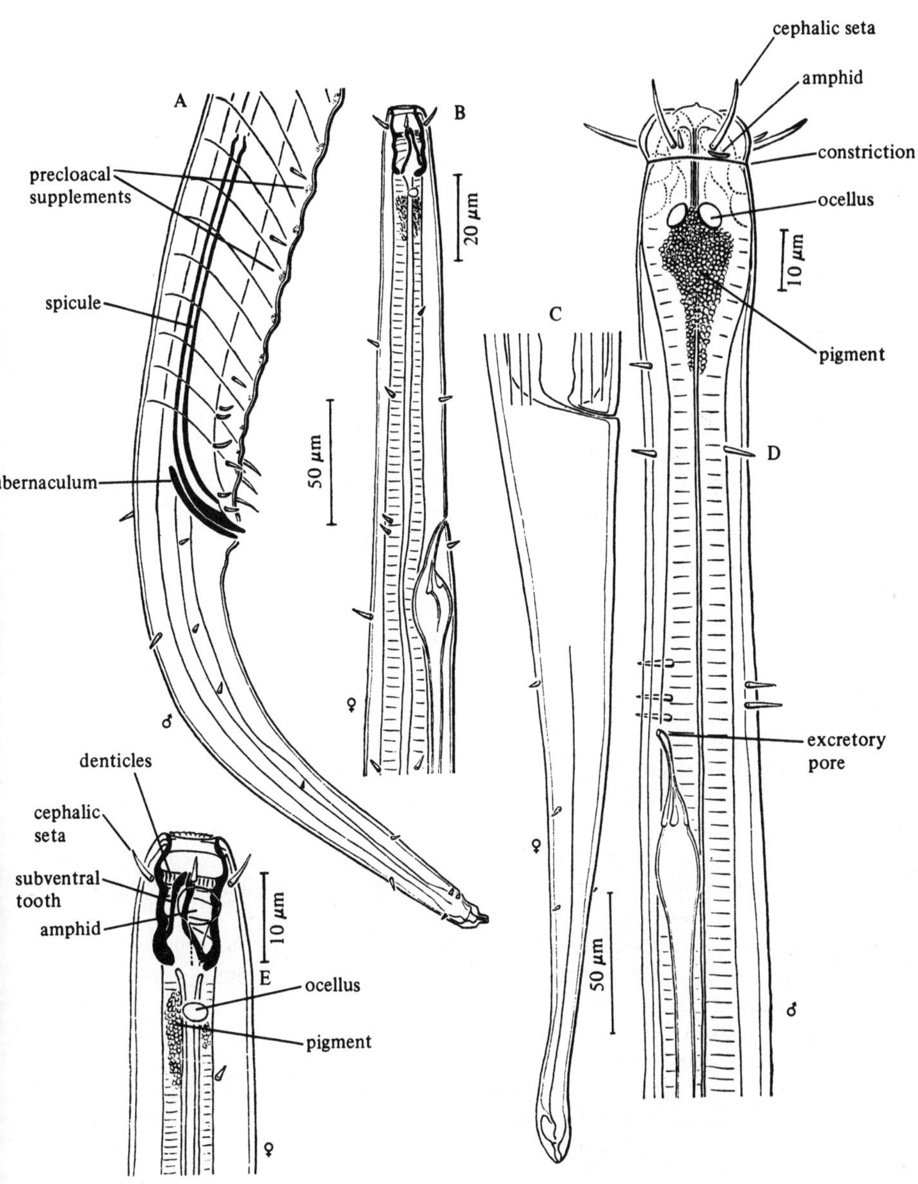

Fig. 120. *Symplocostoma tenuicolle*. A, Male tail; B, Anterior end of female; C, Female tail; D, Anterior end of male; E, Female head (from Luc and De Coninck, 1959).

Genus BATHYLAIMUS Cobb, 1894

Key: Wieser, 1956

Bathylaimus is distinguished from *Tripyloides* by the deeply incised lips and the relatively large buccal cavity, although in *B. stenolaimus* the buccal cavity is almost as small as in some *Tripyloides* species. The structure of the spicules and gubernaculum is rather uniform throughout the genus. The number of teeth on the distal end of the gubernaculum is best determined in ventral view, and this character serves to distinguish *B. stenolaimus* from the other species. Characters used to separate the species are the length and form of the cephalic setae, the size and position of the amphids and the shape of the tail. Many older descriptions show the cephalic setae unjointed, but this feature has probably been overlooked and it is likely that all species have jointed setae. The buccal cavity varies considerably in appearance depending on the angle from which it is viewed; this is particularly true of the tooth-like structures in the posterior section.

A further species, *B. filicaudatus* (Stekhoven and Adam, 1931) has been reported from the Blyth estuary (Northumberland) and the Isles of Scilly, but it is known only from juveniles and is of doubtful validity.

Species: 31

Bathylaimus australis Cobb, 1894
(Fig. 121)

Description. Body length 2.0–2.2 mm. Maximum diameter 40–69 μm (a = 30–50). Cuticle smooth. Mouth surrounded by three high rounded lips which are deeply incised. Labial setae 4 μm, conical. Six long four-jointed cephalic setae 17–21 μm long (0.8 h.d.); four shorter two-jointed setae half their length. Buccal cavity in two separate sections: the anterior broad, rectangular, heavily cuticularised; the posterior small, weakly cuticularised with a pair of rounded subventral teeth. Amphids circular, 6–7 μm diameter, notched at base where amphidial nerve exits, situated anterior to the constriction between the two sections of the buccal cavity. Oesophagus cylindrical, 0.2 times body length. Nerve ring 0.3–0.35 times oesophagus length from anterior. Tail of male conico-cylindrical, 3.3 a.b.d. long; female tail with apical quarter cylindrical, 4.2 a.b.d. Male tail with longish scattered setae, particularly near tip.

Spicules 42–44 μm long, fairly straight, slightly cephalate proximally.

Gubernaculum 47–48 μm, paired and swollen distally, each half with a rounded lateral projection and a pair of ventro-laterally directed teeth; proximally unpaired with thickened antero-ventral rib and narrow ala.

Ovaries small, paired, opposed, reflexed. Vulva at 49–52% of body length.

Distribution. Exe estuary; Strangford Lough, Northern Ireland (intertidal muddy sands); Isles of Scilly (among *Corallina*).

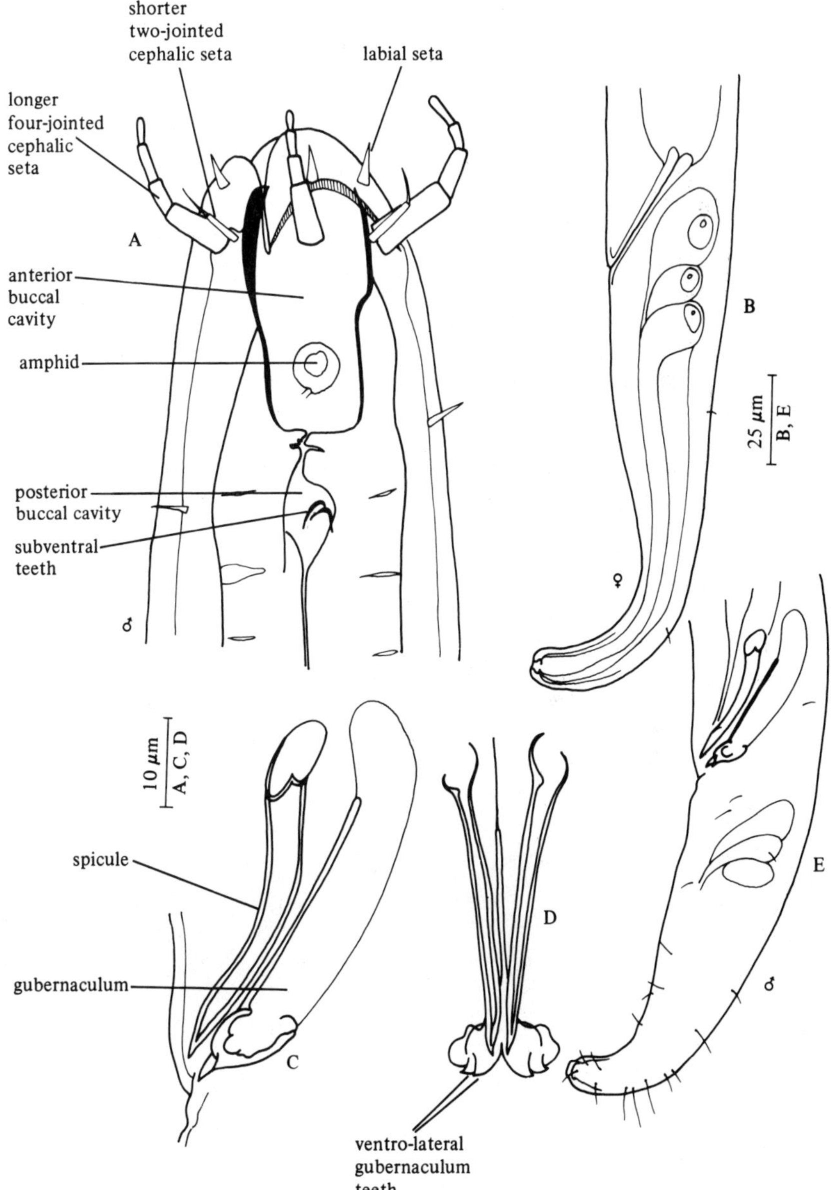

Fig. 121. *Bathylaimus australis*. A, Male head; B, Female tail; C, Spicules and gubernaculum (lateral); D, Spicules and gubernaculum (ventral); E, Male tail. Original.

Bathylaimus capacosus Hopper, 1962
(Fig. 122)

Description. Body length 1.0–1.7 mm. Maximum diameter 29–52 μm (a = 29–49). Cuticle smooth. Setae in cervical region up to 8 μm long, increase abruptly to 21 μm at about 0.4 times the oesophagus length from the anterior. In vulvar region of female two subventral rows of 8 or 9 subventral setae up to 28 μm, denser than those on remainder of body. Male with a pair of very long subterminal setae on tail, up to half the tail length. Labial setae three-jointed, long, 0.5 h.d. Longer cephalic setae four-jointed, 1.3–1.4 h.d. Four shorter setae very short, unjointed. Amphids subspiral, 0.18 times c.d. wide in female, 0.3 in male; level with or behind the posterior section of the buccal cavity.

Spicules and gubernaculum similar to *B. australis*, both 24–45 μm long. Vulva at 57–62% of body length.

Distribution. Loch Ewe, Scotland (intertidal sand).

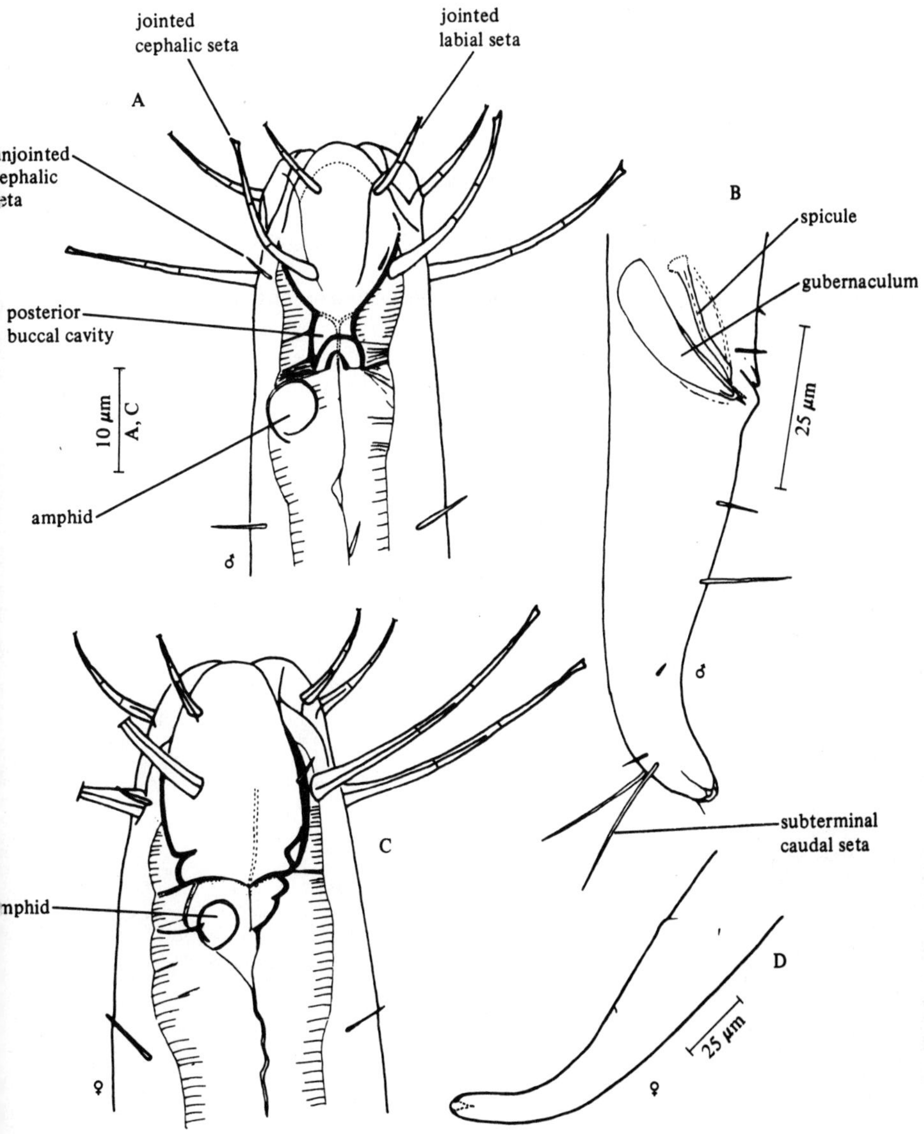

Fig. 122. *Bathylaimus capacosus*. A, Male head; B, Male tail; C, Female head; D, Female tail (from Riemann, 1966).

Bathylaimus inermis (Ditlevsen, 1918)
(Fig. 123)

Description. Body length 2.7–3.8 mm (a = 23–42). Cuticle smooth. Labial setae two-jointed, 0.2 h.d. long. Longer cephalic setae four-jointed, 0.64 h.d.; shorter setae two-jointed, 0.27 h.d. Amphids subspiral, 0.23 times c.d. in diameter, situated at base of buccal cavity. Tail 3 a.b.d. long with a *distinctly swollen tip* in both sexes.

Spicules and gubernaculum similar to preceding species.

Distribution. Exe estuary (intertidal sand or low water mark).

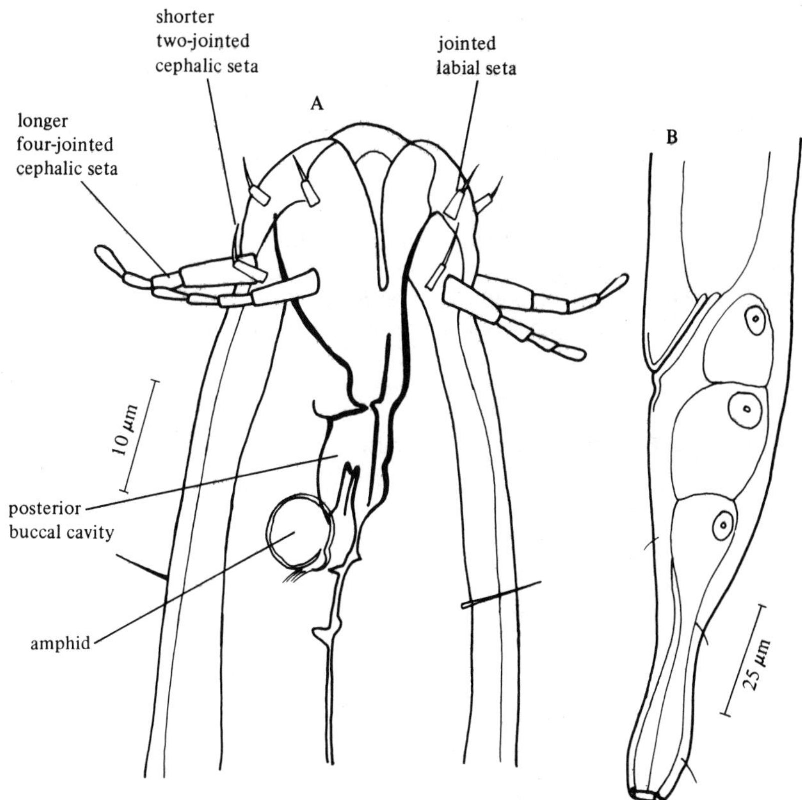

Fig. 123. *Bathylaimus inermis*. A, Juvenile head; B, Juvenile tail. Original.

Bathylaimus paralongisetosus Stekhoven and De Coninck, 1933
(Fig. 124)

Description. Body length 2.2–2.3 mm. Maximum diameter 32–49 μm (a = 46–71). Cuticle with fine transverse striations. Lips high, rounded, striated on outer margins. Labial setae two-jointed, 0.58 h.d. long. Long cephalic setae 1.5 h.d., four-jointed; four shorter setae 0.2 h.d., unjointed. Amphids subspiral, 0.40–0.58 times c.d. in males, 0.26 times c.d. in females; level with posterior section of buccal cavity. Tail 3.2–4.1 a.b.d. long, conical in male but with distal half cylindrical in female. Male tail with two files of ventrolateral setae, and paired 13 μm terminal setae.

Spicules 35–37 μm, gubernaculum same length; both similar in form to *B. australis.*

Vulva at 62% of body length.

Distribution. Exe estuary (intertidal sand).

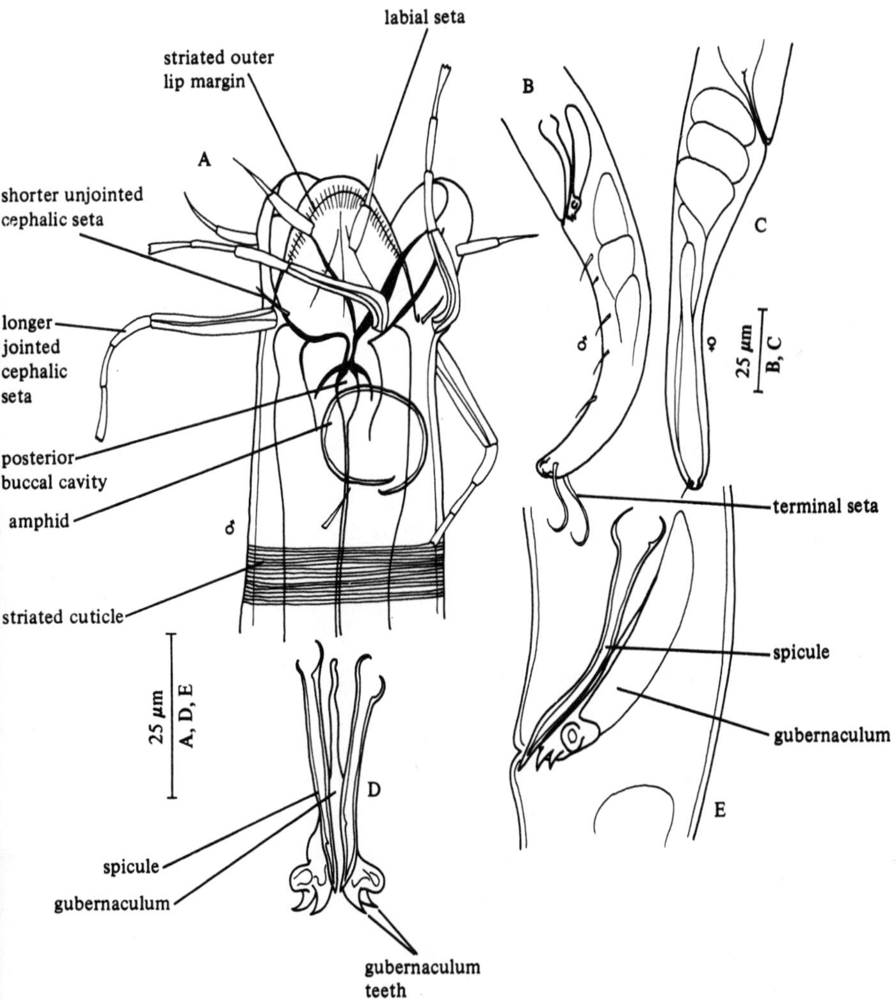

Fig. 124. *Bathylaimus paralongisetosus*. A, Male head; B, Male tail; C, Female tail; D, Spicules and gubernaculum (ventral); E, Spicules and gubernaculum (lateral). Original.

Bathylaimus stenolaimus Stekhoven and De Coninck, 1933
(Fig. 125)

Description. Body length 1.9–2.4 mm. Maximum diameter 33–46 μm (a = 43–62). Cuticle with fine transverse striations. Labial setae two-jointed, 0.2 h.d. long. Six longer cephalic setae four-jointed, 1.2 h.d.; shorter setae two-jointed, 0.55 h.d. Buccal cavity relatively smaller than in other species. Amphids describe a spiral of 1.5 turns, 0.23 times c.d. in diameter, level with base of buccal cavity. Tail 3.2 a.b.d. long, finger-like in shape in both sexes.

Spicules 29–31 μm long.

Gubernaculum 25–26 μm. Each half of gubernaculum has only a single tooth distally.

Vulva at 44–50% of body length.

Distribution. Exe estuary; Northern Ireland; Loch Ewe, Scotland; Isles of Scilly (intertidal and shallow subtidal sand).

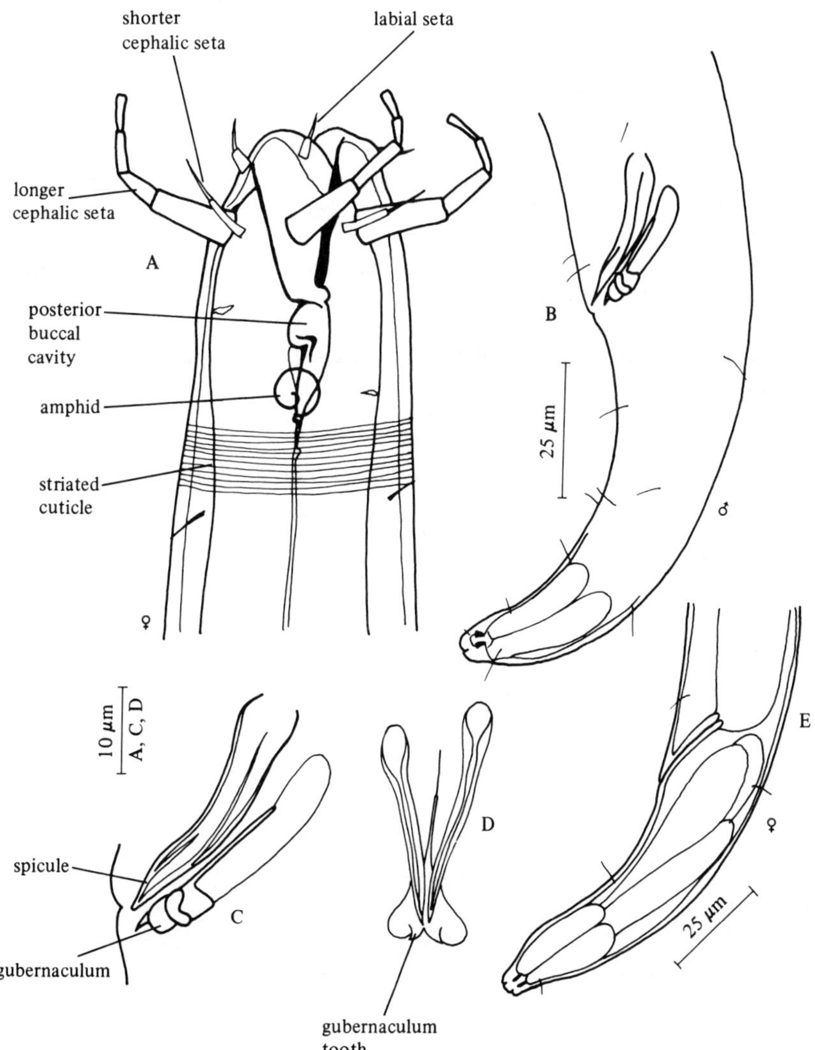

Fig. 125. *Bathylaimus stenolaimus*. A, Female head; B, Male tail; C, Spicules and gubernaculum (lateral); D, Spicules and gubernaculum (ventral); E, Female tail. Original.

Bathylaimus tenuicaudatus (Allgén, 1933)
(Fig. 126)

Description. Body length 1.6–2.1 mm. Maximum diameter 34–56 μm (a = 33–48). Very fine transverse striation of cuticle sometimes discernible. Lips very high, rather pointed, striated on outer margins. Six labial setae two-jointed, 0.45 h.d. long. Longer cephalic setae four-jointed, 1.4–1.6 h.d.; shorter setae unjointed, 0.3 h.d. Amphids subspiral, much larger in male (0.5 times c.d.) than female (0.2 times c.d.); situated at base of buccal cavity. Tail 5.7–6.6 a.b.d. long. Numerous circumcloacal setae in male and two ventro-lateral files of setae down tail length.

Spicules 29–30 μm long.

Gubernaculum 30 μm, similar to *B. australis* (i.e. each half of gubernaculum with two distal teeth).

Vulva at 54–62% of body length.

Distribution. Exe estuary; Loch Etive, Scotland (intertidal sand).

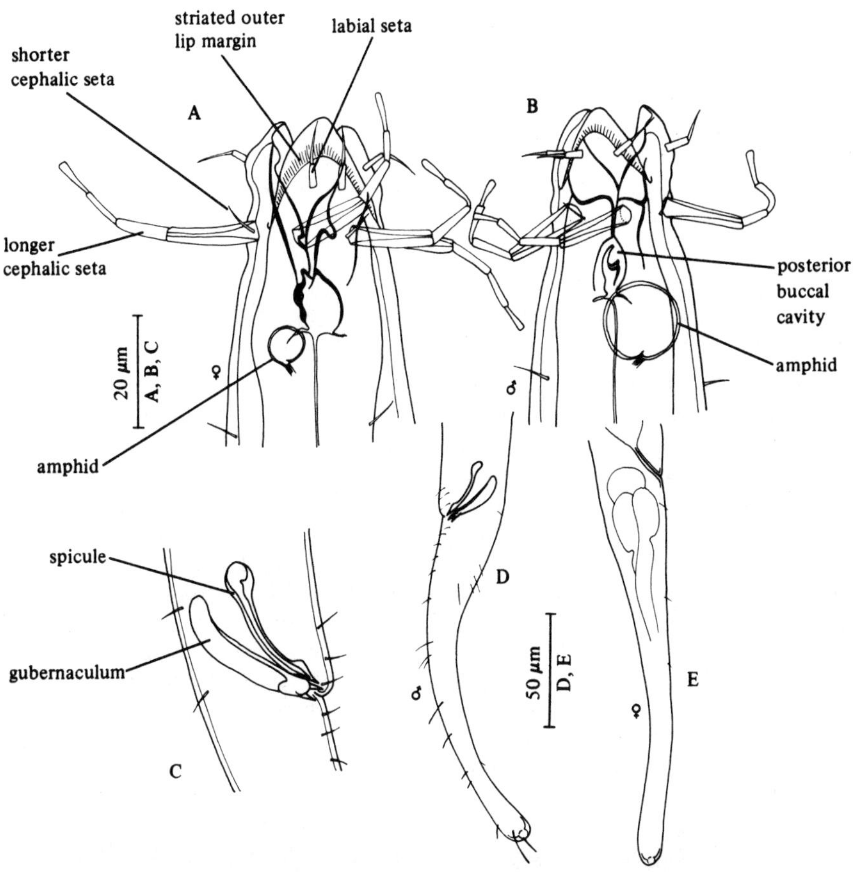

Fig. 126. *Bathylaimus tenuicaudatus*. A, Female head; B, Male head; C, Spicules and gubernaculum; D, Male tail; E, Female tail. Original.

Genus GAIRLEANEMA Warwick and Platt, 1973

Gairleanema has a spherical buccal cavity with a strong dorsal tooth, labial setae with bulbous swellings, and mamilliform precloacal supplements in the male. The amphids are difficult to see: Warwick and Platt (1973) interpreted a small pocket-like structure just behind the lateral cephalic setae as the amphids, but Lorenzen (1981a) considered that an indistinct single-looped structure more posteriorly was the amphid.

Species : 1

Gairleanema anagremilae Warwick and Platt, 1973
(Fig. 127)

Description. Body length 1.9–2.0 mm. Maximum diameter 41–44 μm (a = 45–46). Cuticle with rather indistinct transverse striation. Mouth surrounded by three high lips each with a pair of 28–30 μm labial setae. The labial setae have bulbous swelling about half-way along their length. Ten cephalic setae; six longer ones 68–69 μm with a thin section near the tip; four shorter 7 μm. There are a few short (up to 4.5 μm) setae scattered on the anterior cervical region. Caudal setae and two subventral 17 μm terminal setae present, but other somatic setation lacking. Buccal cavity with a pointed, heavily cuticularised, dorsal onchium pierced by a duct which leads into the oesophageal musculature. Other less distinct structures in the buccal cavity have been interpreted as two weakly cuticularised subventral teeth with rounded tips. Position and structure of amphids equivocal; there is a small indistinct pocket-like structure just behind the lateral cephalic setae with the point of junction of what may be the amphidial nerve prominent; also an indistinct single-looped structure about 11.5 μm wide more posteriorly, the inner margin of the loop being most prominent. Oesophagus cylindrical, 0.175 times body length, 22–25 μm wide at its base. Nerve ring 144–150 μm from anterior. Tail 3.9–4.8 a.b.d. long, posterior half cylindrical.

Spicules 52–53 μm (1.4 a.b.d.) long, equal, L-shaped, cephalate proximally and with broad ventral alae.

Gubernaculum paired, 21 μm long, with hooked dorsal apophyses. There is a ventral precloacal seta 11.5–13 μm long, a small ventral papilla 12.5–13 μm in front of the cloaca and two large rounded mamilliform protuberances in front of this. The posterior protuberance is 32 μm in front of the cloaca and the anterior 43–48 μm in front of the cloaca. The nipples are pierced by fine pores.

Distribution. Loch Ewe, Scotland (intertidal sand).

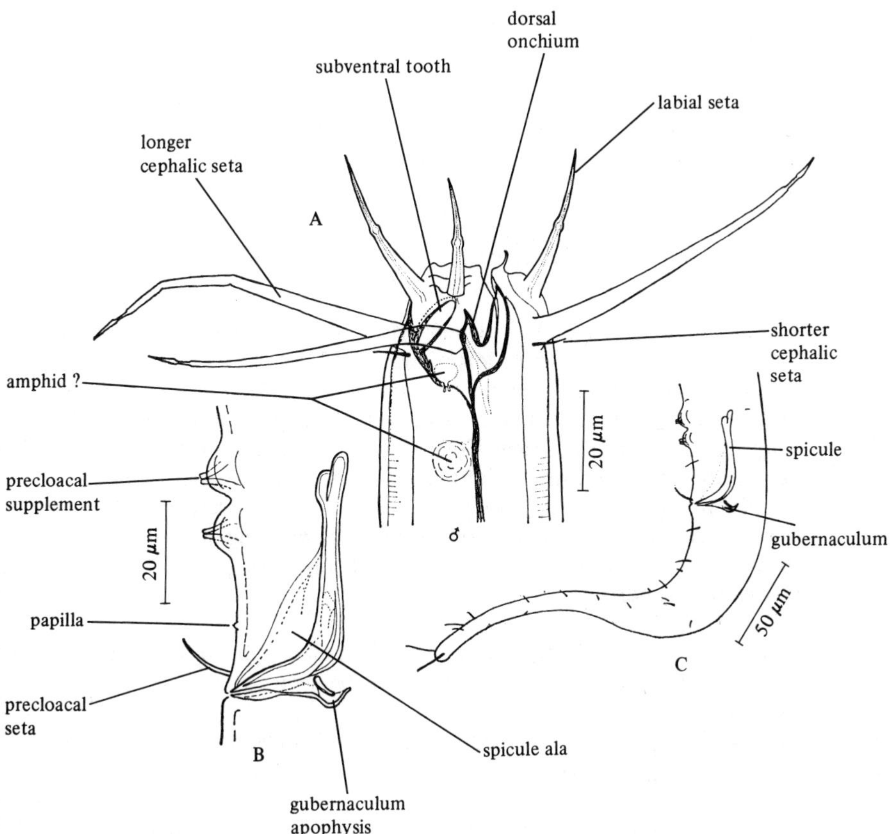

Fig. 127. *Gairleanema anagremilae*. A, Male head; B, Cloacal region; C, Male tail. Original.

Genus TRIPYLOIDES De Man, 1886

Key: Wieser, 1956

Tripyloides is distinguished from *Bathylaimus* by the low lips which are not separated by deep incisions, and the smaller buccal cavity. The two British species are easily separated by the number of chambers of the buccal cavity.

Species: 9

Tripyloides gracilis (Ditlevsen, 1918)
(Fig. 128)

Description. Body length 1.7–2.0 mm. Maximum diameter 42–58 μm (a = 32–42). Cuticle smooth. Three low rounded lips, not deeply incised. Labial setae small, conical. Six longer cephalic setae 8–9 μm (0.5 h.d.), two-jointed; shorter four 5–6 μm, unjointed. Buccal cavity in two sections: anterior section conical, heavily cuticularised, divided in middle by fine cuticular ridge; posterior section small with two teeth. Amphids circular, notched at base, 6–7 μm in diameter, posterior to buccal cavity. Oesophagus more or less cylindrical 0.14 times body length. Nerve ring 0.41–0.51 times oesophagus length from anterior. Tail 3.7–4.5 a.b.d. long, with distal third to a quarter cylindrical.

Spicules 31–32 μm.

Gubernaculum 32–34 μm, similar to *Bathylaimus*, with two pointed teeth on each distal half, the more lateral tooth being much smaller than the median one.

Ovaries paired, symmetrical, opposed, reflexed. Vulva at 51–52% of body length.

Distribution. Skippers Island, Essex; Exe estuary (intertidal mud and muddy sand).

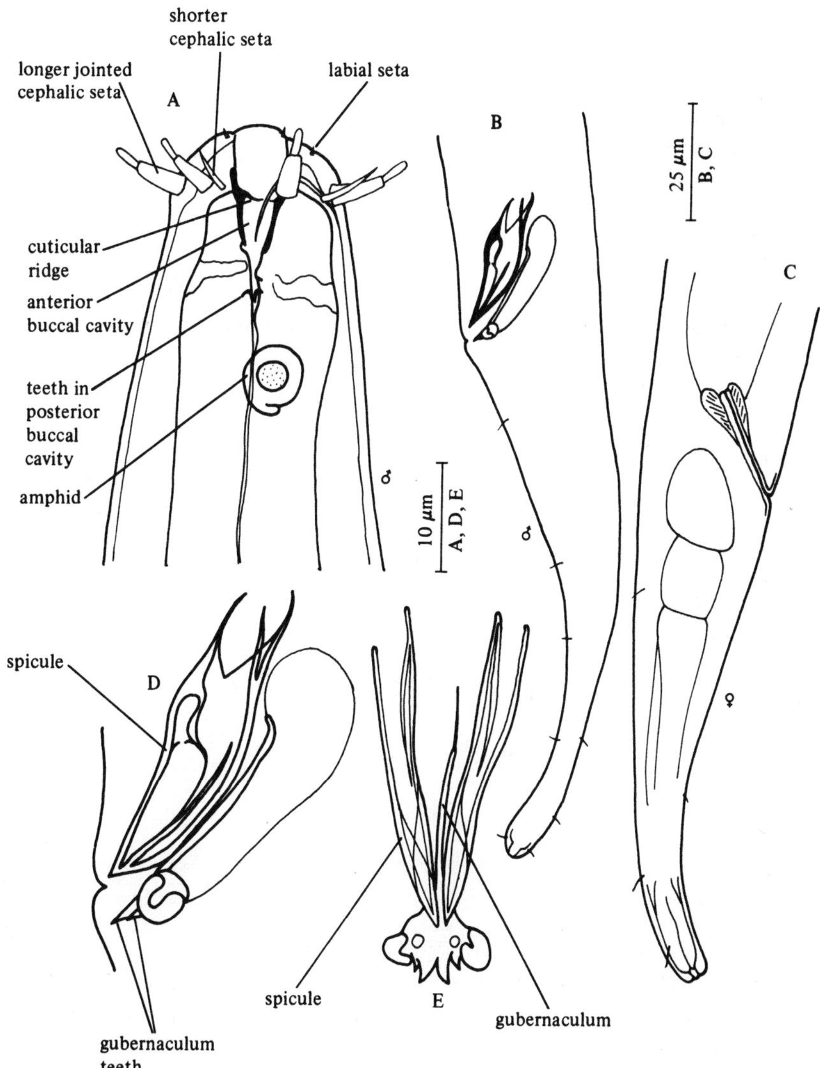

Fig. 128. *Tripyloides gracilis*. A, Male head; B, Male tail; C, Female tail; D, Spicules and gubernaculum (lateral); E, Spicules and gubernaculum (ventral). Original.

Tripyloides marinus (Bütschli, 1874)
(Fig. 129)

Description. Body length 1.9–2.6 mm (a = 34–45). Cuticle smooth. Six longer cephalic setae 6 μm, two-jointed; shorter four 4 μm, unjointed. Buccal cavity in *four* separate sections, fairly heavily cuticularised, posteriormost cavity with two subventral teeth. Amphids circular, 5 μm diameter, level with base of buccal cavity. Tail 4.3 a.b.d. long.

Spicules 23 μm.

Gubernaculum about the same length, similar in form to *T. gracilis.*

Distribution. Blyth estuary; Exe estuary (intertidal mud and sand).

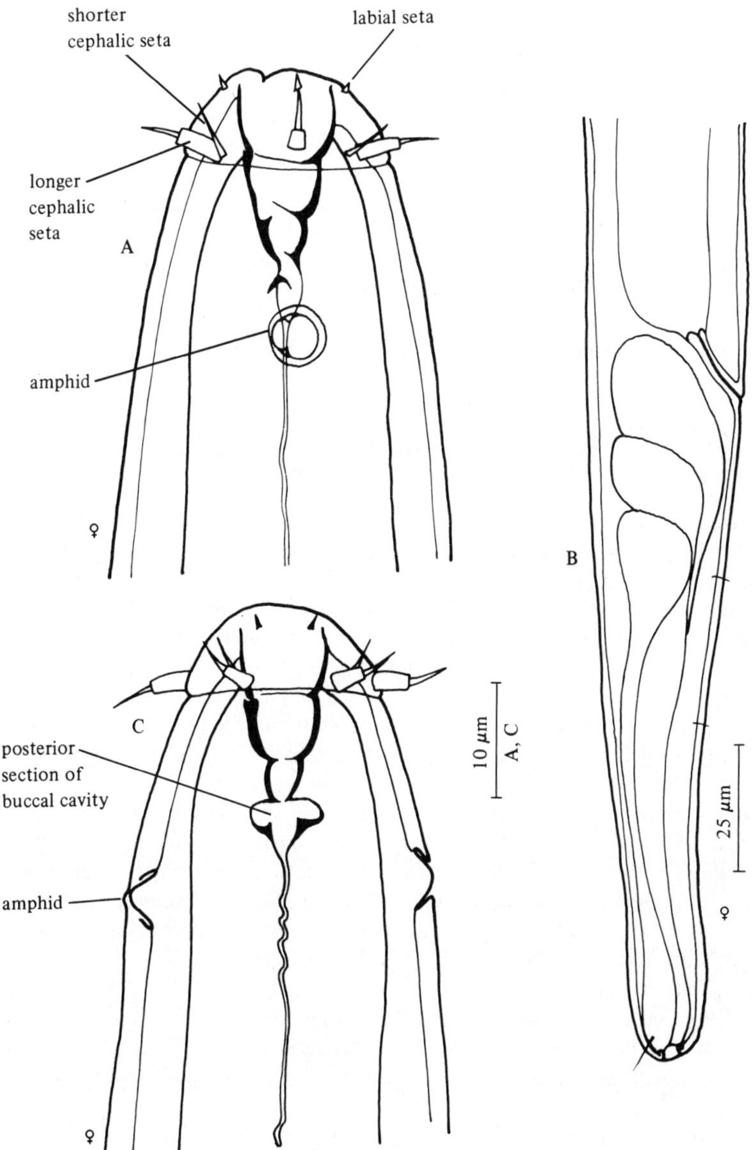

Fig. 129. *Tripyloides marinus*. A, Female head (lateral); B, Female tail; C, Female head (dorsal). Original.

Genus RHABDODEMANIA Baylis and Daubney, 1926

Keys: Boucher, 1971; Platonova, 1976

Rhabdodemania species usually have three teeth at the base of the buccal cavity, but they may be absent, as in *R. imer*. Species in the genus are distinguished primarily on a combination of the length and relative position of the two circles of cephalic setae, length of spicule and length of tail.

R. major and *R. minor* are rather alike in general appearance and structure, but *R. minor* is shorter and stouter and the tail is proportionately larger. *R. imer* is much smaller and more elongated than the other two known British species, with cephalic setae longer than one head diameter: the only other species with such long cephalic setae are *R. coronata* Gerlach, 1952 and *R. illgi* Wieser, 1959, both of which have teeth in the buccal cavity.

Species: 20

Rhabdodemania major (Southern, 1914)
(Fig. 130)

Description. Large species; body length 7.5–8.2 mm. Maximum diameter 97–110 μm (a = 76–81). Cuticle smooth, devoid of body setae except for fine sublateral setae in oesophageal region and tail. Head sharply narrows level with cephalic setae. Six minute papillae at anterior end. Four anterior cephalic setae 9 μm and six posterior cephalic setae 20–21 μm (0.6 h.d.) long. Amphids are faint sinusoidal structures beginning at a small pore 10–15 μm behind the lateral cephalic setae and extending 200–225 μm from the anterior, ending in a horse-shoe shaped structure. The amphid is most tightly looped in the anterior part and gradually begins to straighten out further back. Buccal cavity conical, bearing anteriorly 2 pairs of subdorsal cuticularised teeth and posteriorly a large dorsal and two subventral teeth. Oesophagus tissue surrounds the posterior half of the buccal cavity. The oesophagus has no bulb, but widens posterior to the nerve ring. Nerve ring 0.4 times oesophagus length from anterior. Excretory pore level with base of buccal cavity. Tail cylindrical with complex spinneret and four stout terminal setae; 2.1–2.5 a.b.d. Two caudal glands.

Spicules 80 μm (0.9 a.b.d.) long.

Gubernaculum 46 μm and distally lies lateral to spicule tips. About 40 minute ventral supplements extend about 2200 μm anterior to cloaca. Fine subventral setae present anterior and posterior to cloaca. One outstretched testis.

Vulva at 60% of body length. Two opposed, reflexed ovaries.

Distribution. West Ireland (sand and shells at 44 m); Isles of Scilly (intertidal coarse sand).

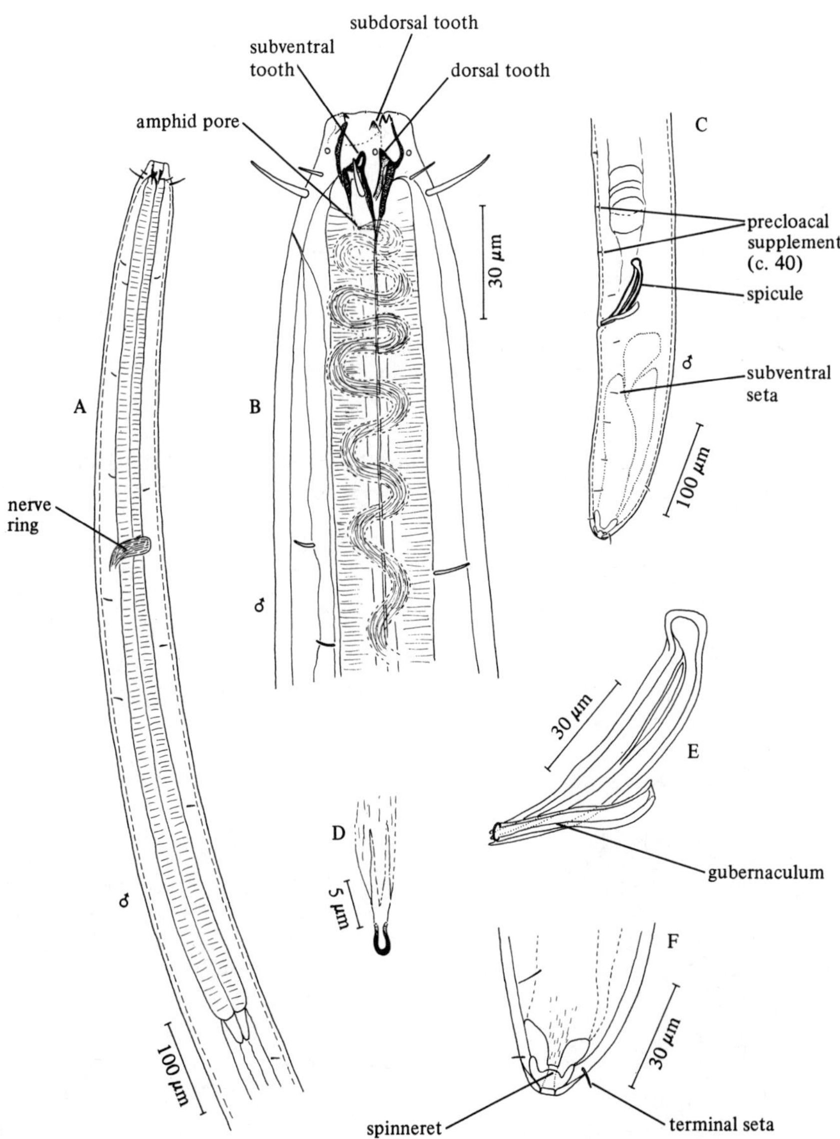

Fig. 130. *Rhabdodemania major*. A, Anterior end of male; B, Male head; C, Male tail; D, Posterior end of amphid; E, Spicules and gubernaculum; F, Tail tip. Original.

Rhabdodemania minor (Southern, 1914)
(Fig. 131)

Description. Body length 4.8–5.0 mm. Maximum diameter 86 μm (a = 56–58). Cuticle smooth with a few somatic setae anterior to nerve ring and on tail but otherwise absent. Four anterior cephalic setae 5 μm long, situated nearer the six 12 μm (0.5 h.d.) cephalic setae than in *R. major*. Amphids extend 180 μm (0.57 times distance from anterior to nerve ring) but the anterior origin has not been observed. Excretory pore opens slightly behind the base of the buccal cavity. Nerve ring 0.46 times oesophagus length from anterior. Tail cylindrical, 2.8–2.9 a.b.d.

Spicules 58 μm (0.8 a.b.d.) long.

Gubernaculum 43 μm. 18 minute precloacal supplements.

Vulva at 58% of body length. Two opposed, reflexed ovaries.

Distribution. West Ireland (at 26–48 m); Loch Etive, Scotland (shallow subtidal sand); Isles of Scilly (fine to coarse sand, intertidal to 30 m).

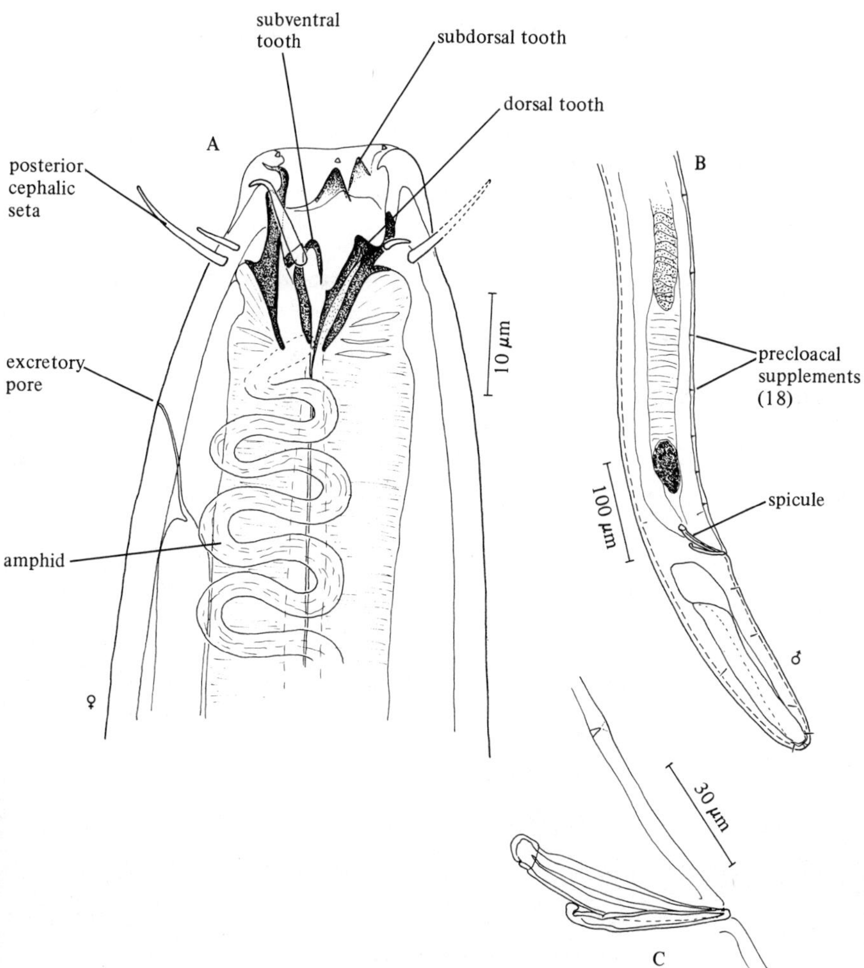

Fig. 131. *Rhabdodemania minor*. A, Female head; B, Posterior end of male; C, Spicules and gubernaculum. Original.

Rhabdodemania imer Warwick and Platt, 1973
(Fig. 132)

Description. Body length 3.0 mm. Maximum diameter 24 μm (a = 123). Cuticle smooth. Four 4 μm cephalic setae situated anterior to six 16.5 μm (1.3 h.d.) cephalic setae. Amphids and excretory pore not seen. Buccal cavity conical, teeth absent. Nerve ring 0.46 times oesophagus length from anterior. Tail cylindrical, 5.8 a.b.d.

Spicules 26 μm (1.1 a.b.d.) long.

Gubernaculum 14 μm. 6 minute supplements extending 147 μm anterior to cloaca.

Female not described.

Distribution. Loch Ewe, Scotland (intertidal sand).

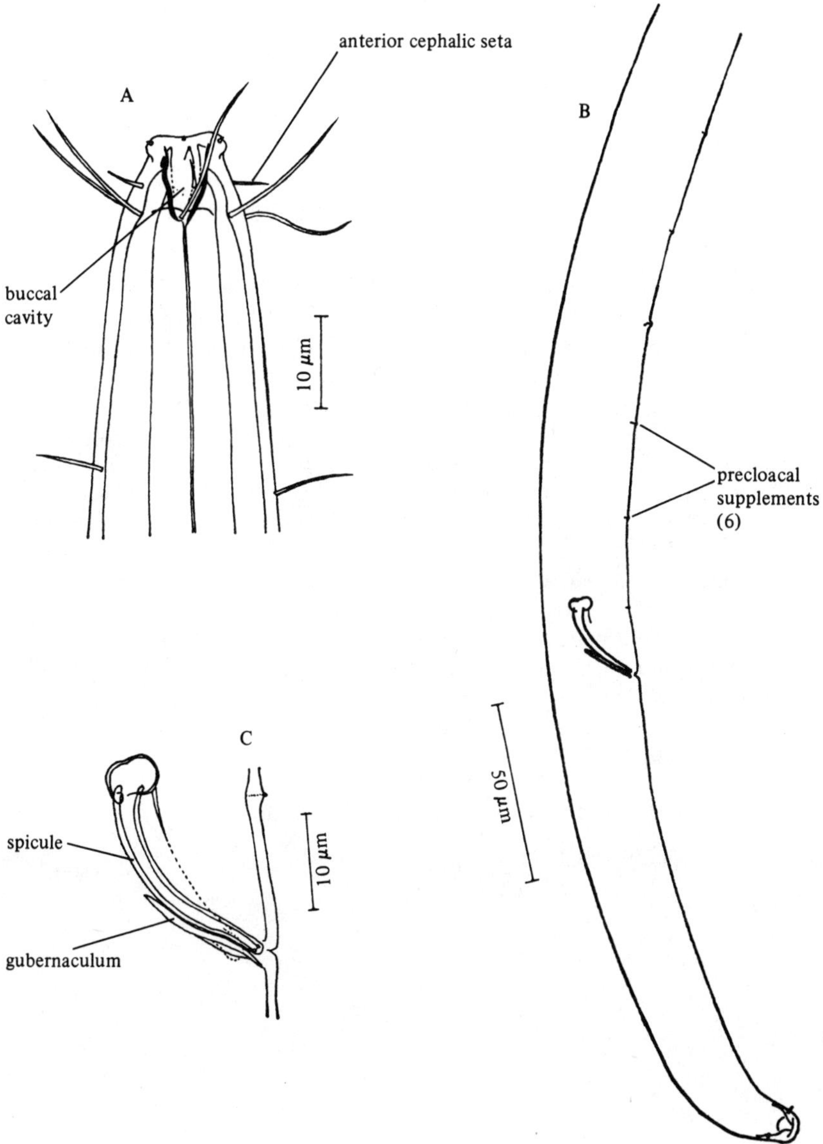

Fig. 132. *Rhabdodemania imer*. A, Male head; B, Posterior end of male; C, Spicules and gubernaculum. Original.

Genus RHABDOCOMA Cobb, 1920

Rhabdocoma is distinguished from the next genus, *Trefusia*, in having a single ovary posterior to the vulva, with the consequent anterior position of the vulva.

Species : 4

Rhabdocoma riemanni Jayasree and Warwick, 1977
(Fig. 133)

Description. Body length 3.3–3.9 mm. Maximum diameter 34–40 μm (a = 91–99). Cuticle smooth. Head bears three lips each having two small labial papillae 1 μm long. Buccal cavity very small, 3.5 μm deep and conical in shape. Six cephalic setae, 12–15 μm long, three-jointed. Four submedian cervical setae, 8 μm long, present just anterior to the amphids. Amphids circular in outline, but appear to be deeper and layered dorsally, 9 μm diameter (0.42 times c.d.), situated 25–28 μm behind anterior end. Oesophageal bulb absent. Tail very long and filamentous and constitutes approximately one-fifth to two-fifths of the total body length, 660–1370 μm (22–46 a.b.d.) long in males and 855–1340 μm (26–40 a.b.d.) long in females.

Spicules paired, equal, 31 μm long, slightly cephalated proximally and pointed distally. A prominent median ridge present in both spicules.

Gubernaculum paired, very narrow, lies parallel to the spicules, 11 μm long. 9–13 precloacal supplements present. These are in the form of small conical papillae each bearing a fine filamentous seta through their centres.

Ovary single and postvulvar. Vulva very narrow and indistinct, at 24–38% of body length.

Distribution. Firth of Clyde (beach sand below mean tide level).

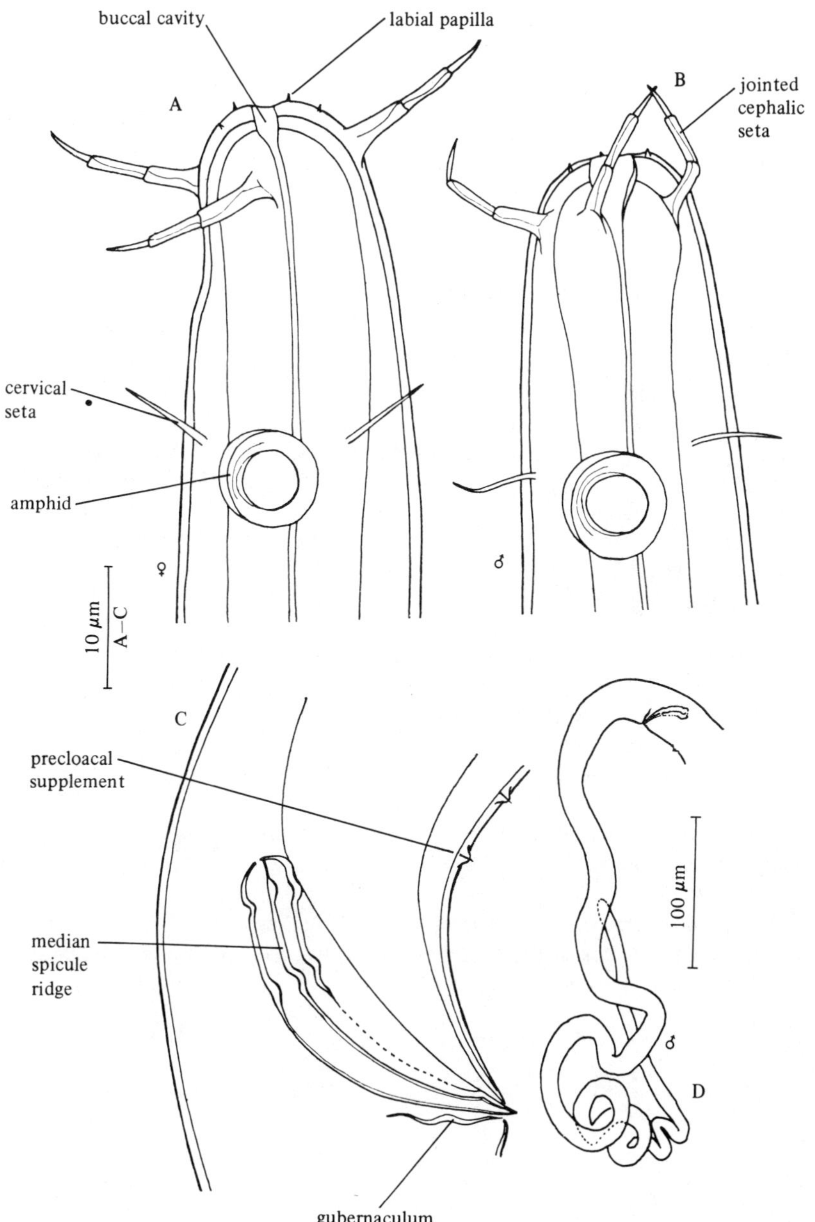

Fig. 133. *Rhabdocoma riemanni*. A, Female head; B, Male head; C, Cloacal region; D, Male tail (from Jayasree and Warwick, 1977).

Genus TREFUSIA De Man, 1893

Key: Riemann, 1966
Trefusia species have jointed cephalic setae, no buccal cavity and (unlike *Rhabdocoma*), two opposed ovaries. Species are distinguished on the relative lengths and position of the head setae, the length of the spicules, the form of the amphids and the length of the tail.

Species : 10

Trefusia longicaudata De Man, 1893
(Fig. 134)

Description. Body length 1.9–2.6 mm. Maximum diameter 26–33 μm (a = 58–80). Cuticle with fine transverse striations. Six rounded lips each with a small labial papilla. Six cephalic setae 7–11 μm, each with stout basal section and filiform tip. Four cervical setae 4–6 μm, level with or slightly posterior to amphids. Buccal cavity small, conical. Amphids pocket-like, 3–5 μm wide. Oesophagus about 0.1 times body length, cylindrical throughout. Just anterior to the amphids a ring of six prominent conical strands join the oesophagus to the cuticle of the body wall. Nerve ring at about 0.5 times oesophagus length from anterior. Tail extremely long and filiform, up to 40 a.b.d.

Spicules 24–27 μm, with dorsal alae and small central wedges near the proximal ends.

Gubernaculum 15–16 μm, paired, notched distally to form two minute teeth. No supplements.

Ovaries paired, reflexed, posterior one slightly larger than anterior. Vulva at about 36% of body length in undamaged specimens, but as the tail tip frequently gets broken off during extraction and processing, this measurement is not reliable.

Distribution. Recorded from intertidal and shallow subtidal sediments and holdfasts at several localities around the British Isles, with a preference for muddy sands.

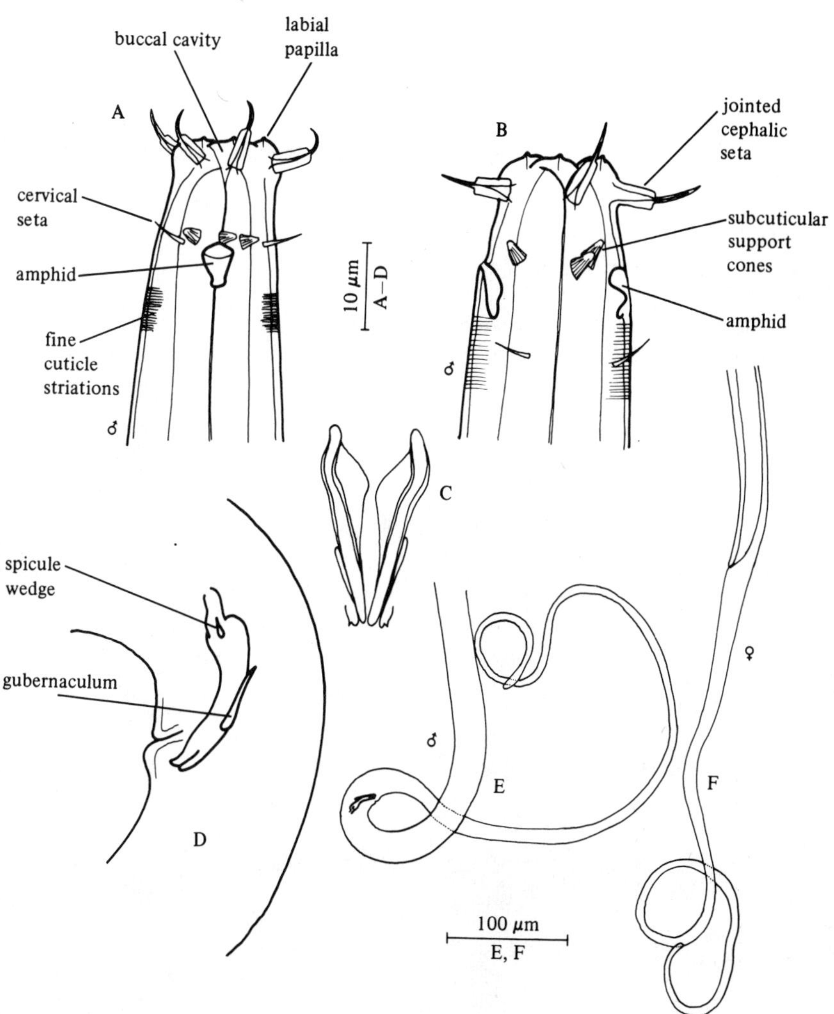

Fig. 134. *Trefusia longicaudata.* A, Male head (lateral); B, Male head (dorsal); C, Spicules and gubernaculum (ventral); D, Cloacal region (lateral); E, Male tail; F, Female tail. Original.

Trefusia zostericola Allgén, 1933
(Fig. 135)

Description. Body length 3.6 mm. Maximum diameter 30 μm (a = 118). Labial setae small, fine, pointed. Six 14 μm cephalic setae three-jointed. Four cervical setae 9 μm, posterior to amphids. Amphids 6 μm wide. Tail 7.3 a.b.d. long.

Spicules 55 μm, arcuate, slightly cephalate proximally.

Gubernaculum a slender rod with two well-developed lateral pieces. Precloacal supplements present.

Distribution. Loch Ewe, Scotland (intertidal sand).

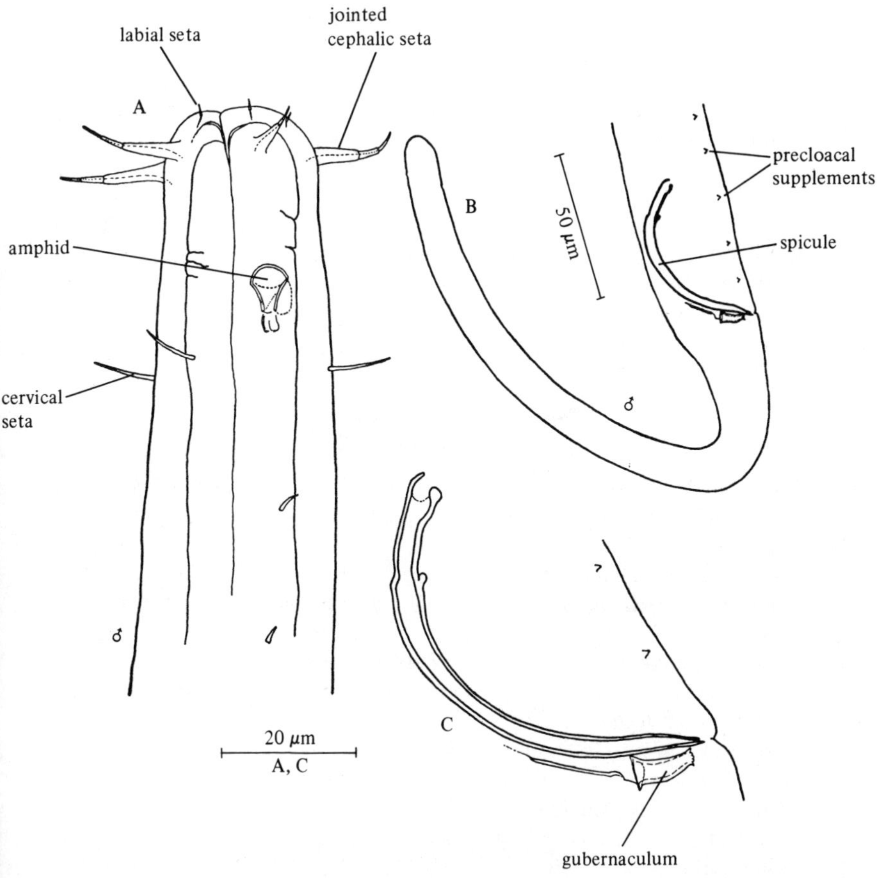

Fig. 135. *Trefusia zostericola.* A, Male head; B, Male tail; C, Spicules and gubernaculum. Original.

Genus RHABDITIS Dujardin, 1845

Review: Inglis and Coles, 1961

This genus is widespread and common on land as freeliving and parasitic forms but only two of the 70 or more known species have ever been recorded from the sea. *R. marina* is quite typical of the strandline in association with rotting material, being a bacterial feeder. Males of the two species are immediately recognisable by the combination of oesophagus shape and the presence of a bursa copulatrix. The only possibility of confusion is with the oesophagus of *Haliplectus* and *Setoplectus* (but these lack a cylindrical buccal cavity and bursa copulatrix) or the bursa copulatrix of *Anoplosoma* and *Oncholaimellus* (different oesophagus shape).

Species : 2 marine

Rhabditis marina Bastian, 1865
(Fig. 136)

Description. Body length 1.4–2.4 mm (a = 13–24). Six labial papillae and ten short cephalic setae in two circles. Amphid a small opening just posterior to cephalic setae. Mouth surrounded by six lobed lips. Buccal cavity cylindrical. Oesophagus characteristic with distinct bulb in the middle as well as posteriorly: length about 0.15 times body length. Tail short and conical.

Spicules 40–70 μm.

Gubernaculum about half length of spicules. Cloaca surrounded by bursa copulatrix (flaps of cuticle) supported by nine pairs of long narrow papillae. Single reflexed testis.

Two reflexed ovaries. Vulva at 50–54% of body length.

Distribution. Reported from many locations around the coast of the British Isles; found intertidally in sand, weeds and especially in association with rotting seaweed.

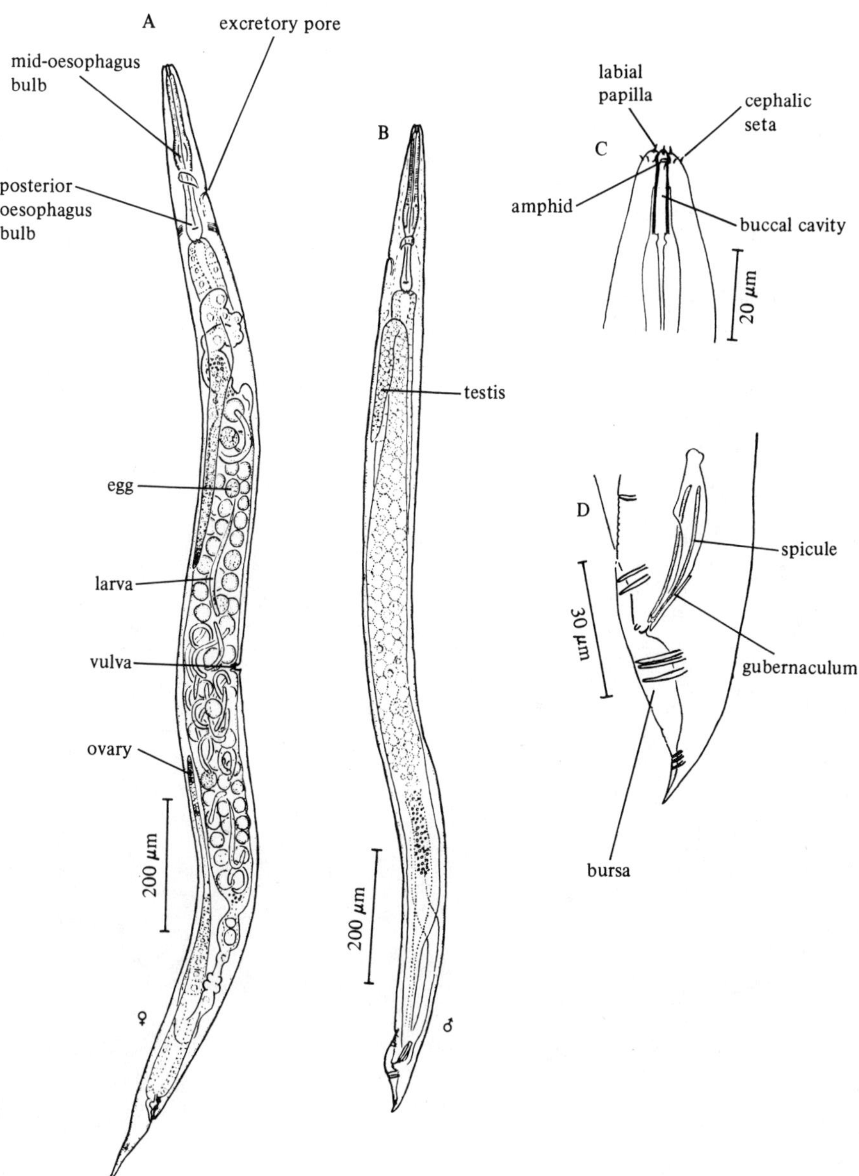

Fig. 136. *Rhabditis marina*. A, Entire female containing eggs and larvae; B, Entire male; C, Head; D, Male tail (from Inglis and Coles, 1961).

Rhabditis ehrenbaumi Bresslau and Stekhoven, 1935
(Fig. 137)

Description. Body length 0.9–1.6 mm (a = 14–21). Head similar to *R. marina.* Oesophagus with characteristic middle and posterior bulbs; length about 25% of total body length. Tail short and rounded with a pointed terminal spike.

Spicules about 70 μm.

Gubernaculum about one-third of spicule length. Cloaca surrounded by bursa copulatrix consisting of flaps of cuticle supported by ten pairs of papillae. Single reflexed testis.

Two reflexed ovaries. Vulva at 52–60% of body length.

Distribution. Found in rotting seaweed in South Wales, Plymouth and Weston-super-Mare but occurrences may only be accidental introductions from the land.

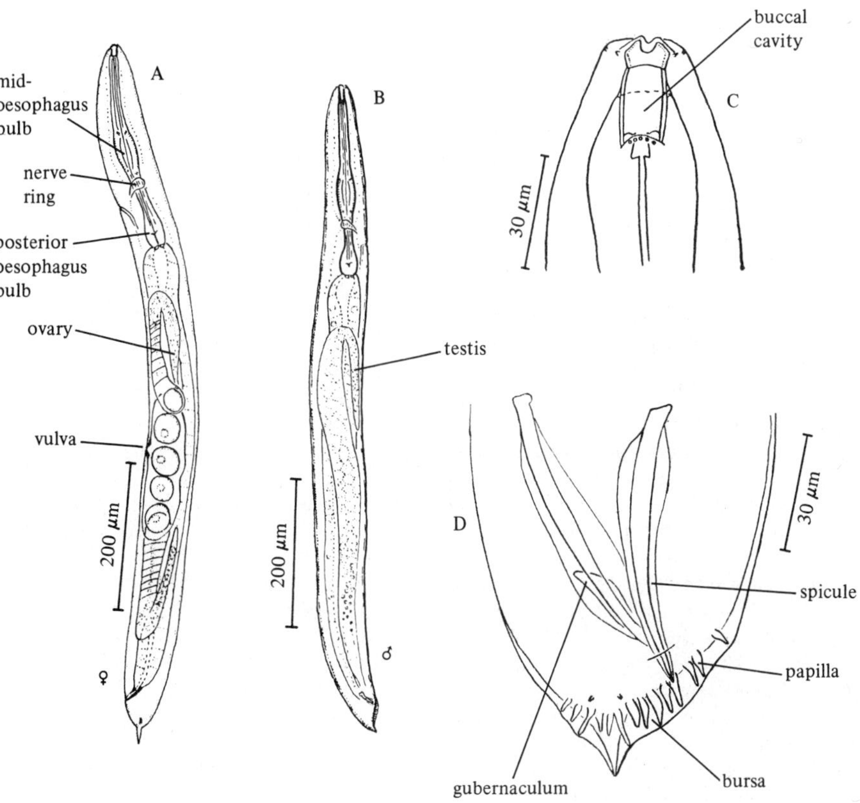

Fig. 137. *Rhabditis ehrenbaumi*. A, Entire female; B, Entire male; C, Head; D, Male tail (from Inglis and Coles, 1961).

Glossary of Special Terms

ala (*pl.* alae) Longitudinal cuticle ridge or thickening forming a winglike extension. Also a thin winglike extension of the spicule.

amphid Paired lateral sense organs situated on or just posterior to the head.

annulation Conspicuous transverse cuticle striation.

annule The transverse section between striae.

apophysis (*pl.* apophyses) A separate process off the main structure, especially off the gubernaculum.

buccal cavity Mouth.

bulb Muscular swelling of the oesophagus.

bursa (*pl.* bursae) Winglike extension of the cuticle surrounding the cloaca in the male.

cardia Muscular structure at the posterior end of the oesophagus connecting to and often within the anterior part of the intestine.

caudal Belonging to the tail.

caudal glands Unicellular glands, usually three, lying within or sometimes extending anterior to tail but discharging through a spinneret at the tail tip.

cephalic Belonging to the head.

cephalic capsule Modified, usually strongly cuticularised, inner layer of cuticle at the point where the anterior part of the oesophagus attaches to the body wall. Often the posterior limit is marked by an external groove or suture.

cephalic incision Anterior extensions of the cephalic suture which divide the posterior part of the capsule into lobes.

cephalic ventricle Fluid-filled space anterior to the end of the oesophagus.

cervical Belonging to the neck, which extends from the base of the head to the posterior end of the oesophagus.

clavate Club shaped.

cloaca The chamber in the male into which the products of both the digestive and reproductive systems enter to be voided via the cloacal opening. In the female, only the digestive system empties at this position so that the opening is an anus.

corpus gelatum Gelatinous substance filling the cavity of the amphid.

cuticle Outer covering of the body which also lines the buccal cavity, oesophagus lumen, vagina and rectum.

cuticularised Formed of cuticle but may also be used more generally to denote any structure which appears to be rigid or strengthened relative to surrounding tissue.

demanian system Complex seminal receptical (sperm store) found in females of the family Oncholaimidae.

denticle Small tooth or projection, often found in groups or bands.

desmen Conspicuous transverse rings around the bodies of desmoscolecids.

didelphic Having two ovaries.

dioecious Having separate males and females.

diorchic Having two testes.

distal Furthest from the point of origin. Closest to the exterior when referring to parts of the spicule.
ductus entericus Part of the demanian system between osmium and uvette.
ductus uterinus The connection between the demanian system and the uterus.
excretory pore Ventral opening in the cuticle through which the products of the excretory system are voided.
fenestra (*pl.* fenestrae) Rounded area or window at the anterior end of the cephalic incision which often accommodates the base of a cephalic seta.
filiform Shaped like a thread.
fovea Cavity of the amphid.
fusiform Shaped like a spindle, tapering at both ends.
gonad An ovary or testis.
gubernaculum A cuticularised guiding piece lying dorsal to the spicules in the cloaca.
incision See cephalic incision.
intestine Simple tube composed of a single layer of cells responsible for the digestive process. Gut.
isthmus A narrow section of the oesophagus.
juvenile Any of the four immature stages before the adult stage.
labial Belonging to the lips.
labium (*pl.* labia) Lip.
lacuna A gap.
lamella (*pl.* lamellae) A thin sheet or sheetlike process.
lateral chord Longitudinal thickening of the hypodermis in the lateral region.
lateral differentiation Elaboration or ornamentation of the cuticle confined to the lateral region.
lip The anteriormost part of the head cuticle, usually divided into three or six lobes.
loxometaneme Metaneme arranged at an angle to the longitudinal body axis.
lumen Cavity contained by the walls of a tube, sac or cell.
mandible Cuticularised moveable structure in the buccal cavity used for gripping or biting.
mandibular ring Transverse bar in the buccal cavity formed from extensions of the mandibles.
meiofauna Small multicellular organisms able to pass through a 1 mm sieve, normally found in benthic sediments or seaweeds.
metaneme Filamentous stretch receptors found in the lateral epidermal region.
metazoa Multicellular organisms.
moniliform glands Bead-like cells, usually in the form of a rosette, surrounding the uvette in the demanian system.
monodelphic Having one ovary.
monorchic Having one testis.
nerve ring Discrete ring of nerve tissue around the oesophagus.
ocellus (*pl.* ocelli) Light sensitive pigmented body containing a lens-like structure.
odontium (*pl.* odontia) A tooth which is formed in the oesophagus but moves forward to the anterior of the buccal cavity to become functional.
oesophagus Muscular tube connecting the buccal cavity to the intestine. Pharynx.
onchial plate Base of the onchium.
onchium (*pl.* onchia) A tooth which is formed in the buccal cavity; in some enoplids attached to the mandible.
orthometaneme Metaneme arranged parallel to the longitudinal body axis.
osmium Modified intestinal epithelium connected to the demanian system.

papilla (*pl.* papillae) Small nipple-like projection of the cuticle.
papilliform Shaped like a papilla or nipple.
peduncle Stem or stalk, often the base for a seta.
pharyngeal bulb Muscular swelling of the oesophageal tissue around the buccal cavity.
pharynx Often used in nematology to refer to the buccal cavity but in strict anatomical terms synonymous with the oesophagus.
plaque Thickened patch on the cuticle; often forms the base for the amphid in certain species.
proximal Nearest to the point of origin. Furthest from the exterior when referring to parts of the spicule.
pseudocoelom The body cavity.
punctation Dotted cuticle, formed from tiny raised knobs, rods or rounded depressions in the cuticle.
radial masses Structures in the buccal cavity linking the mandibular ring to the radial processes.
radial processes Structures supporting the onchial plate.
rectum Part of the alimentary canal connecting the intestine to the anus in females or cloaca in males, usually seen as a flattened tube.
reflexed Folded back on itself, especially of testes and ovaries.
renette cell The ventral excretory gland, unicellular.
reniform Shaped like a kidney.
rostrum Non-annulated part of the head of draconemas.
scapulus Central part of a metaneme.
sensillum (*pl.* sensilla) A general term for a sense organ.
seta (*pl.* setae) Relatively elongated hair-like sense organ or sensillum.
setiform Shaped like a hair or bristle.
somatic Belonging to the body; often used to refer to sensilla on the general body surface.
spinneret Terminal pore of the caudal glands.
spicule (*pl.* spicules) Copulatory organs, usually paired, in the cloaca of the male.
stoma Mouth.
stria (*pl.* striae) Transverse groove in the cuticle.
striated The condition where the cuticle bears a series of striae, often very close together.
stylet A long slender spear-like structure in the buccal cavity.
subcephalic setae Setae located on or just behind the head which appear to be in some way associated with the cephalic setae, often found as a circle of six or in six groups.
subsidiary lobe Distinct anterior extension of the lip-lobe.
supplements In males, additional copulatory organs usually situated ventrally or subventrally anterior to the cloaca and often appearing to have a secretory function.
tail The portion of the body posterior to the cloaca or anus.
tropis Hollow tooth-like structure formed by the ventral wall of the cephalic capsule in certain enoplids.
turgor Distension of cell or body by fluid contents.
uvette Swollen structure at the junction of the main duct of the demanian system and the ductus uterinus.
vas deferens Tube connecting the testis to the ejaculatory duct.
vermiform Shaped like a worm.
vulva Opening of the female reproductive system.

Literature list

NOTE: References to all the taxonomic literature prior to 1972 may be found in Gerlach and Riemann (1973/1974). Serious students of marine nematodes will find this work essential.

Allen, M. W. and Noffsinger, E. M. 1978. *A revision of the marine nematodes of the Superfamily Draconematoidea Filipjev, 1918 (Nematoda: Draconematina).* University of California Press, Berkeley. 133 pp.

Andrássy, I. 1973. Über vier homonym Nematodengattungen. *Nematologica*, **19,** 403–4.

Andrássy, I. 1976. *Evolution as a basis for the systematization of nematodes.* Pitman Publishing, London. 288 pp.

Benwell, M. P. 1981. A new species of *Gonionchus* (Nematoda: Xyalidae) from the Firth of Clyde, with a redescription of *Enoploides spiculohamatus* Schulz (Nematoda: Enoplidae). *Cah. Biol. mar.*, **22,** 177–84.

Belogurov, O. I. and Belogurova, L. S. 1980. Morphology of *Belbolla intarma* sp. n., diagnosis and table for species determination of the genus *Belbolla. Biologiya Morya*, **4,** 74–7 (in Russian).

Blome, D. 1974. Zur Systematik von Nematoden aus dem Sandstrand der Nordseeinsel Sylt. *Mikrofauna Meeresbodens*, **33,** 1–25.

Boucher, G. 1975. Nématodes des sables fins infralittoraux de la Pierre Noire (Manche occidentale). I. Desmodorida. *Bull. Mus. natn. Hist. nat., Paris*, **285,** 101–28.

Boucher, G. and De Bovée, F. 1972. *Trochamus carinatus* gen. et sp. n. et *Adeuchromadora megamphida* gen. et sp. n. Chromadoridae (Nematoda) à dix soies céphaliques de la vase terrigène côtière de Banyuls-sur-Mer. *Vie Milieu*, **22,** 231–42.

Boucher, G. and Helléouët, M.-N. 1977. Nématodes des sables fins infralittoraux de la Pierre Noire (Manche occidentale). III. Araeolaimida et Monhysterida. *Bull. Mus. natn. Hist. nat. Paris*, **427,** 85–122.

Freudenhammer, I. 1975. Desmoscolecida aus der Iberischen tiefsee, zugleich eine Revision dieser Nematoden-Ordnung. *Meteor ForschErgebn.*, **20,** 1–65.

Gerlach, S. A. and Riemann, F. 1973/1974. The Bremerhaven checklist of aquatic nematodes. A catalogue of Nematoda Adenophorea excluding the Dorylaimida. *Veröff. Inst. Meeresforsch. Bremerh.*, Suppl. 4, Part 1 (1973) and Part 2 (1974), 1–736.

Gerlach, S. A. and Schrage, M. 1972. Life cycles at low temperatures in some free-living marine nematodes. *Veröff. Inst. Meeresforsch. Bremerh.*, **14,** 5–11.

Gourbault, N. 1980. Nématodes abyssaux (Campagne Walda du N/O "J. Charcot"). II. Espèces et genre nouveaux de Comesomatidae. *Bull. Mus. natn. Hist. nat. Paris*, **3,** 737–49.

Haspeslach, G. 1973. Superfamille des Ceramonematoidea (Cobb, 1933) (Nematoda), évolution et systématique. *Annals Soc. r. zool. Belg.*, **102,** 235–51.

Heip, C., Smol, N. and Absillis, V. 1978. Influence of temperature on the reproductive potential of *Oncholaimus oxyuris* (Nematoda: Oncholaimidae). *Mar. Biol.*, **45,** 255–60.

Heip, C., Vincx, M., Smol, N. and Vranken, G. 1982. The systematics and ecology of free-living marine nematodes. *Helminth. Abstr.* (Series B), **51,** 1–31.

Hopper, B. E. 1977. *Marylynnia*, a new name for *Marilynia* of Hopper, 1972. *Zool. Anz.*, **198,** 139–40.

Hopper, B. E., Fell, J. W. and Cefalu, R. C. 1973. Effect of temperature on life cycles of nematodes associated with the mangrove (*Rhizophora mangle*) detrital system. *Mar. Biol.*, **23,** 293–6.

Jayasree, K. and Warwick, R. M. 1977. Free-living marine nematodes of a polluted sandy beach in the Firth of Clyde, Scotland. Description of seven new species. *J. nat. Hist.*, **11,** 289–302.

Jensen, P. 1978. Revision of Microlaimidae, erection of Molgolaimidae fam. n., and remarks on the systematic position of *Paramicrolaimus* (Nematoda, Desmodorida). *Zoologica Scr.*, **7,** 159–73.

Juario, J. V. 1974. Neue freilebende Nematoden aus dem Sublitoral der Deutschen Bucht. *Veröff. Inst. Meeresforsch. Bremerh.*, **14,** 275–303.

Lambshead, P. J. D. and Platt, H. M. 1979. *Bathyeurystomina*, a new genus of freeliving marine nematodes (Enchelidiidae) from the Rockall Trough. *Cah. Biol. mar.*, **20,** 371–80.

Lopez, G., Riemann, F. and Schrage, M. 1979. Feeding biology of the brackish-water Oncholaimid nematode *Adoncholaimus thalassophygas. Mar. Biol.*, **54,** 311–18.

Lorenzen, S. 1973. Die Familie Epsilonematidae (Nematodes). *Mikrofauna Meeresbodens*, **25,** 1–86.

Lorenzen, S. 1976. *Calomicrolaimus rugatus* n. gen., n. sp. (Desmodoridae, Nematodes) from a sandy beach in Colombia. *Mitt. Inst. Colombo-Aleman Invest. cient. "Punta Betin"*, **8,** 79–82.

Lorenzen, S. 1977. Revision der Xyalidae (freilebende Nematoden) auf der Grundlage einer kritischen Analyse von 56 Arten aus Nord- und Ostsee. *Veröff. Inst. Meeresforsch. Bremerh.*, **16,** 197–261.

Lorenzen, S. 1978a. Postembryonalentwicklung von *Steineria-* und Sphaerolaimidenarten (Nematoden) und ihre Konsequenzen für die Systematik. *Zool. Anz.*, **200,** 53–78.

Lorenzen, S. 1978b. Discovery of stretch receptor organs in nematodes – structure, arrangement and functional analysis. *Zool. Script.*, **7,** 175–8.

Lorenzen, S. 1981a. Entwurf eines phylogenetischen Systems der freilebenden Nematoden. *Veröff. Inst. Meeresforsch. Bremerh.*, Supp. 7, 1–472.

Lorenzen, S. 1981b. Bau, Anordnung und postembryonale Entwicklung von Metanemen bei Nematoden der Ordnung Enoplida. *Veröff. Inst. Meeresforsch. Bremerh.*, **19,** 89–114.

McIntyre, A. D. and Warwick, R. M. (in press). Meiofauna techniques. In *Methods for the Study of Marine Benthos*, 2nd edition, ed. N. A. Holme and A. D. McIntyre, Chapter 7. Blackwells, Oxford.

Platonova, T. A. 1976. Nizshie Enoplida (Svobodnozhivuschie Morskie Nematody) Morei SSSR. In *Nematodes and their role in the meiobenthos. Akad. Nauk CCP, Zool. Inst. Issledovanija Fauny Morjei*, **15**(23), 3–164 (in Russian).

Platt, H. M. 1973. Freeliving marine nematodes from Strangford Lough, Northern Ireland. *Cah. Biol. mar.*, **14,** 295–321.

Platt, H. M. 1982. Revision of the Ethmolaimidae (Nematoda: Chromadorida). *Bull. Br. Mus. nat. Hist.* (*Zool.*), **43** (4), 185–252.

Platt, H. M. and Warwick, R. M. 1980. The significance of freeliving marine nematodes to the littoral ecosystem. In *The shore environment*, Vol. 2: *Ecosystems*, ed. J. H. Price, D. E. G. Irvine and W. F. Farnham, pp. 727–59. Academic Press, London.

Riemann, F. 1974. Trefusialaimus nov. gen. (Nematoda) aus der Iberischen Tiefsee mit Diskussion des männlichen Genitalapparates von Enoplida Tripyloidea. *Meteor ForschErgebn.*, **18**, 39–43.

Schrage, M. and Gerlach, S. A. 1975. Über Greeffiellinae (Nematoda, Desmoscolecida). *Veröff. Inst. Meeresforsch. Bremerh.*, **15**, 37–64.

Smol, N., Heip, C. and Govaert, M. 1980. The life cycle of *Oncholaimus oxuris* (Nematoda) in its habitat. *Ann. Soc. r. Zool. Belg.*, **110**, 87–103.

Warwick, R. M. 1977. Some free-living marine nematodes from the Isles of Scilly. *J. nat. Hist.*, **11**, 381–92.

Warwick, R. M. 1981a. Survival strategies of meiofauna. In *Feeding and survival strategies of estuarine organisms*, ed. N. V. Jones and W. J. Wolff, pp. 39–52, Plenum, New York.

Warwick, R. M. 1981b. The influence of temperature and salinity on energy partitioning in the marine nematode *Diplolaimelloides bruciei. Oecologia*, **51**, 318–25.

Warwick, R. M. and Gage, J. D. 1975. Nearshore zonation of benthic fauna, especially Nematoda, in Loch Etive. *J. mar. biol. Ass. U.K.*, **55**, 295–311.

Warwick, R. M. and Platt, H. M. 1973. New and little known marine nematodes from a Scottish sandy beach. *Cah. Biol. mar.*, **14**, 135–58.

Index to genera and species

The page citations in roman are to the text, those in *italics* are to illustrations.